BH #9450

AF316714

OMIZU SERIES

1987 OIL ON RAG PAPER
60" X 30 1/2"
REMEMBERING FAMILY
VACATIONS IN PUGET SOUND
AND THE FREQUENTLY
CHANGING ASPECT OF THE
OCEAN BLISS TRIED TO
CAPTURE THE FEELING WITH
PURE COLOR. THE SERIES
WAS EXHIBITED AT THE GAYLE
WEYHER GALLERY THAT FALL.

IN MEMORY OF

E. Frank Sanguinetti, 1917 - 2002

George S. Dibble, 1904 - 1992

Katherine Metcalf Nelson

INTERSECTIONS
The Art of Anna Campbell Bliss

May 23 - August 3, 2004

UTAH MUSEUM OF FINE ARTS

University of Utah, Salt Lake City, Utah

For Robert Lewis Bliss

The catalog has been made possible
by a generous grant from the
George S. and Dolores Dorè Eccles Foundation

Co-sponsors:
Dennis & Bonnie Phillips
Noel & Klancy DeNevers
Janet Minden
Bernard Simbari, MD
William Barnett
Joseph & Dorothy Ann Palmer
Wendy Evans Joseph, FAIA
Anonymous
Prescott Muir, AIA
Rita Fordham

Graphic design: Ted Nagata
Text by Katherine Metcalf Nelson
Digital Photography: Skylar Nielsen
Printing by Printers Inc.

Published by Bliss Studio Publication with
the support of the Salt Lake Art Center
on the occasion of the art exhibition at the
Utah Museum of Fine Arts,
University of Utah, Salt Lake City, Utah
Copyright © 2004 by Anna Campbell Bliss

Exhibition Design by David Hardy & A.C. Bliss
Sponsored by the UMFA Special Exhibition Council

Library of Congress Catalog Control Number 2004106076
ISBN 0-9754915-0-4

INTERSECTIONS
The Art of
Anna Campbell Bliss

Mary F. Francey Ph.D.

Curator of American Art
Utah Museum of Fine Arts
Salt Lake City, UT 84112-0350

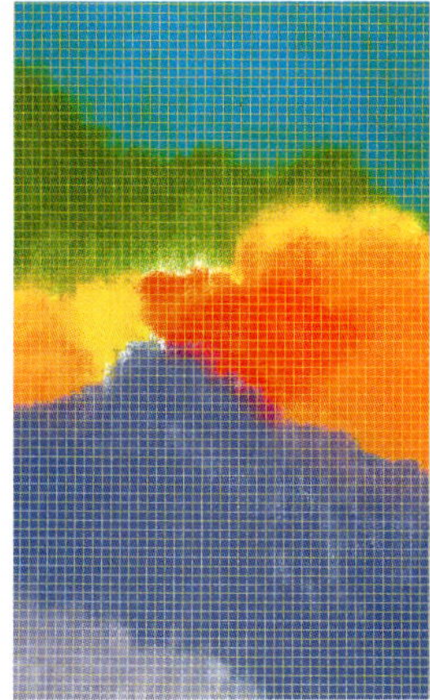

Doors
1992 right panel
32"x 18"
alkyd on enameled steel
collection: Mary Dayton

This exhibition emphasizes the inseparability of art, mathematics and architecture in the work of the artist Anna Campbell Bliss, while revealing the origins and complexity of the creative process. Since she arrived in Utah in 1963, Bliss has raised the level of understanding of contemporary art within the community. Believing that there was a broader base for contemporary art than was apparent, she founded the Contemporary Arts Group in an effort to build a larger audience for local artists. By including science and technology she expanded the base and offered challenges and opportunities for collaboration. Her vision about the interconnectedness of design with the environment has transferred to the community and, to a large extent, increased our awareness of the relationship of objects and images to the spaces they occupy.

An example of how Bliss explores the strong relationships between architectural spaces, mathematical concepts, and artistic expression is WINDOWS (1991), the mural she created for the Data Processing Center in the State Office Building at the Capitol. A pioneering work, WINDOWS is a result of Bliss's explorations of the potential of the computer to produce textures and patterns that could be translated into screen-printed images for installation in a public building. Awarded a fellowship to the American Academy in Rome in 1984, Bliss had ample opportunity to visit and study the great seventeenth century architecture of the city. Baroque churches are,

by definition, a coherent synthesis of painting, sculpture and architecture, a concept that she must have had in mind when designing LIGHT OF GRACE (1993), the 40' by 40' stained glass wall for Saint Thomas More Catholic Church in Sandy, Utah.

Her latest work, EXTENDED VISION (2001-2003) is a mural for the Cowles Mathematics Building, University of Utah, which emphasizes the influence of mathematics in the sciences, art and architecture. This complex work, located in the lobbies and corridors on three floors, encompasses the themes of Numbers and Measures, Intersections, and Outer Space that, once again, affirms her purpose of making connections with the sciences, arts, and the culture from which they emanate.

In addition, Bliss has a strong publication record, with articles appearing in the AIA Journal, Design Quarterly, Art in America, Scientific American, and Leonardo among others. Her recent venture, LABYRINTHS OF THE MIND, is a mixed media artist's book that includes paintings as well as digital and screen prints in a Japanese accordion-fold book in which she explores a range of personal resources. Considered in the aggregate, the contributions that Bliss continues to make to the community cannot be over-stated. Not only is she an effective model for other artists whose work requires thoughtful viewer participation, but her own visual responses to contemporary society are inevitably connected with the history of human intellectual endeavor.

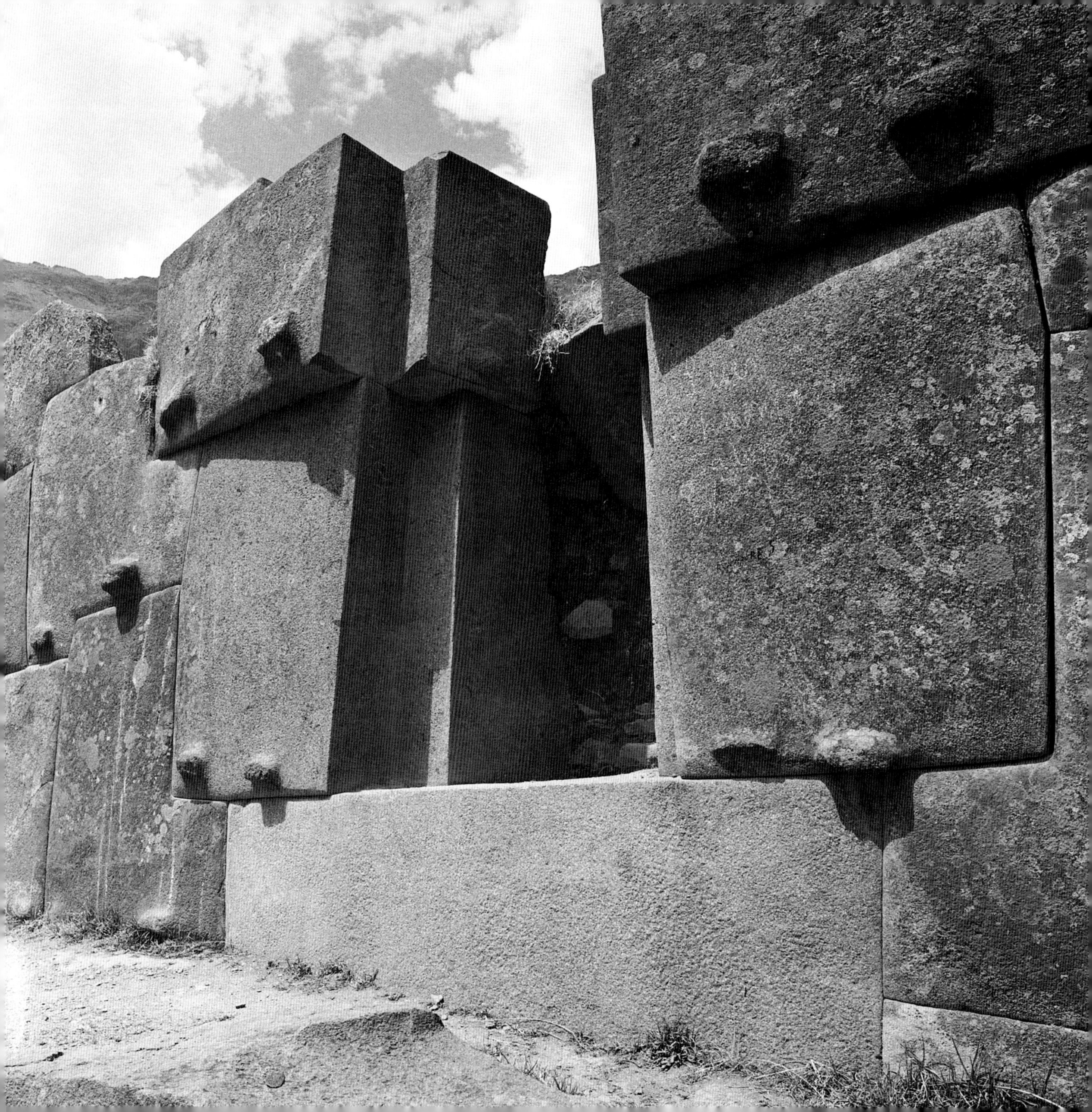

INTERSECTIONS
The Art of
Anna Campbell Bliss

By Katherine Metcalf Nelson

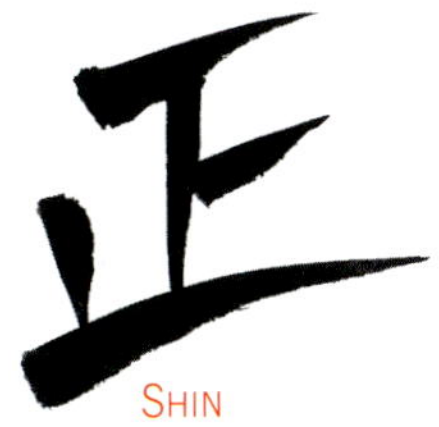

SHIN

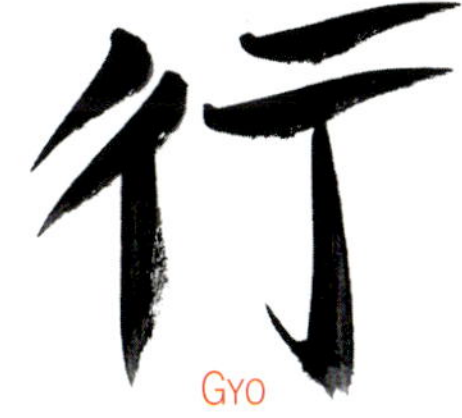

GYO

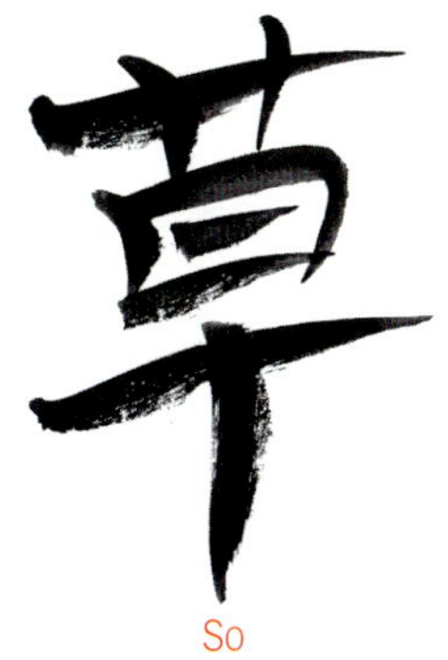

SO

Intersections. This is where we meet, in the passageway between the Data Processing Center and the government chambers of the State Capitol Office Building in Salt Lake City. We are sitting before a shimmering wall of computer generated, screen printed plates of enameled steel. These modules form windows of a giant simulated computer screen, a mural collage of communication systems that is anchored specifically in this time and place and floats way beyond the computer age. Watching people walk by this stabile, flickering installation and listening to the artist speak about her work, I am traveling through her life.

As we talk I sense her passion for archeology and the arts of many cultures; her love of the rhythms of Islamic calligraphy, modern dance and the geodesics of Buckminister Fuller. I can feel her reverence when she speaks of the light in the courtyard of the Villa Giulia or the disciplined intuition of a Japanese Master's ink painting. She is both excited and frustrated in her attempts to merge algorithmic art with eye and hand. All this wonder is woven into her enduring

OPPOSITE: **OLLANTAYTAMBO**, 1969
FROM A PHOTO STUDY IN PERU BY THE ARTIST

fascination with what she calls the intersections of art, science and technology.

While listening I observe vibrations of the changing colors, patterns and layers of WINDOWS, the 8' x 30' mural in front of us. A grid of steel modules becomes a ground for shifting "brushstrokes" that move in waves of light across this field of vision. Monet's Nympheas ripples through me.

For a visiting Japanese architect/scientist observing the work:

"This is shin, gyo, so.
Shin is the formal structure.
Gyo is variation
within the structure.
So is the touch that disturbs
both Shin and Gyo."

Somewhere between Albers' tonal vision and the elegant asymmetry of Japanese design stands the Shin and Gyo of Anna's aesthetic:

Multiple disciplines
in service to structure
in a chromatic composition

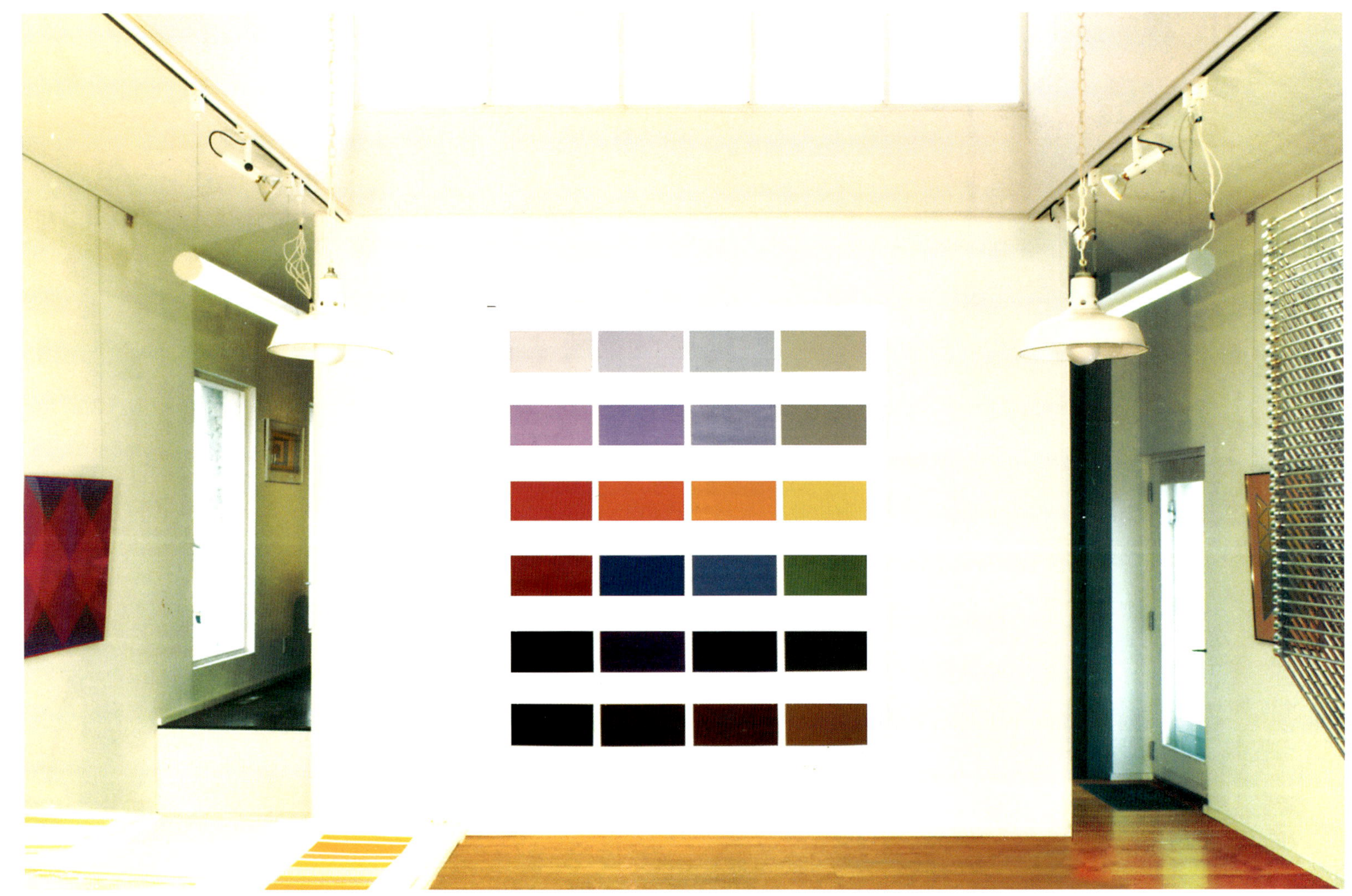

Bliss studio (a remodeled 1898 grocery store)
showing **PALETTE STUDY** for Peerless Lighting Corporation 1986

**Her
navigators:**

Bonnard

Turner

Klee

Constructivists

Matisse

Monet

Kepes

Albers

Buckminster Fuller

Leonardo

Piero della Francesca

**Her instruments
of navigation:**

Graph paper

Oil on canvas

Screen printing

Computer languages

Maps

Photography

Geodesics

Gardens

Italian memories

Islamic calligraphy

Fibonnaci numbers

Vitruvian man

Lake Truchet

Windows

Light of Grace

House of Mathematics

Labyrinths of the Mind

**Beyond the studio, her ideas,
written and spoken, travel
around the world:**

Publications

Lectures

Color consultations

Multi-disciplinary seminars where
ART, SCIENCE, TECHNOLOGY meet

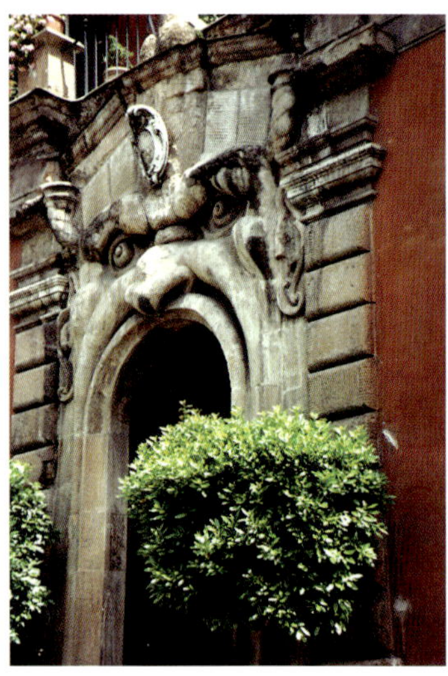

Roman doorway, Bliss photo

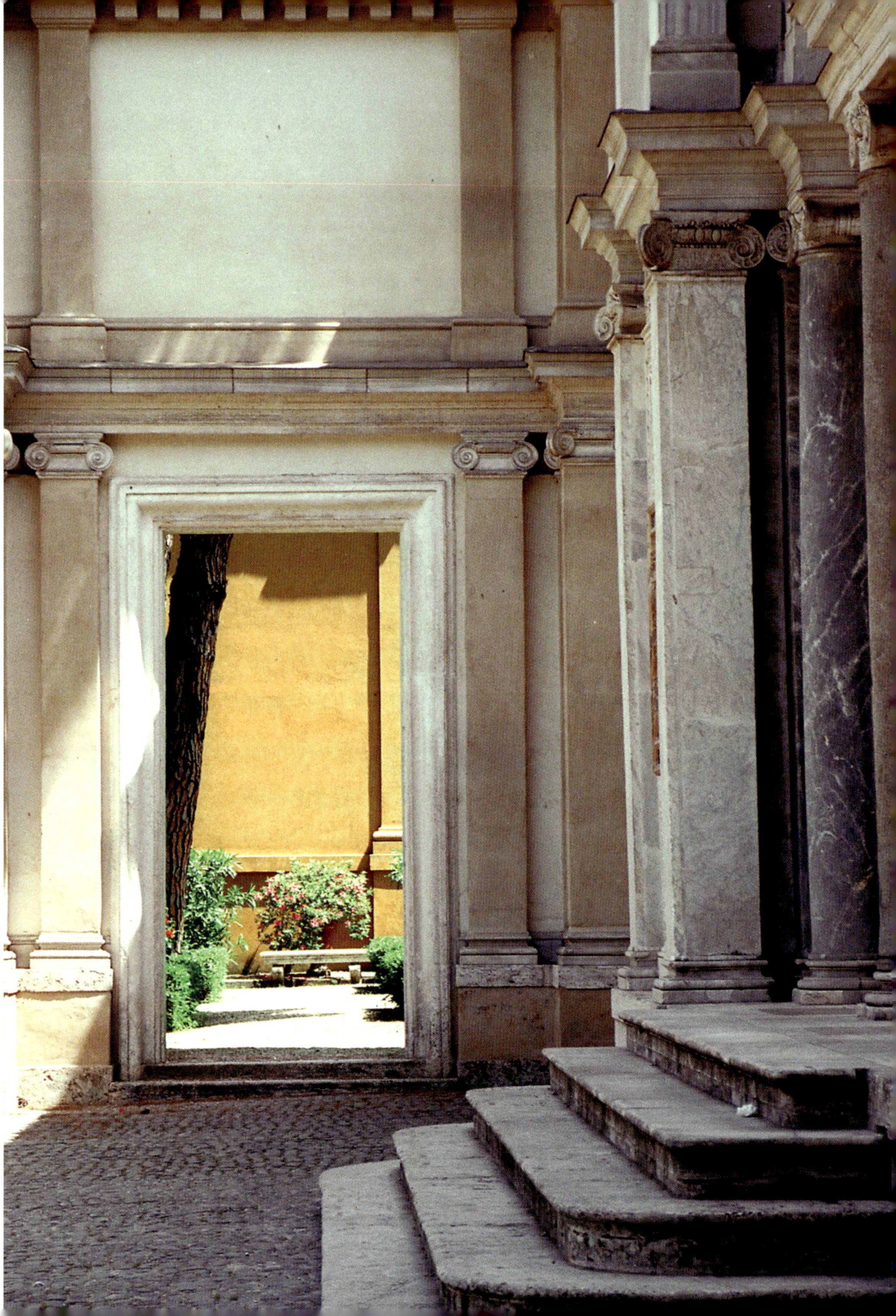

10

VILLA GIULIA
A favorite retreat in Rome.
Photo by the artist.

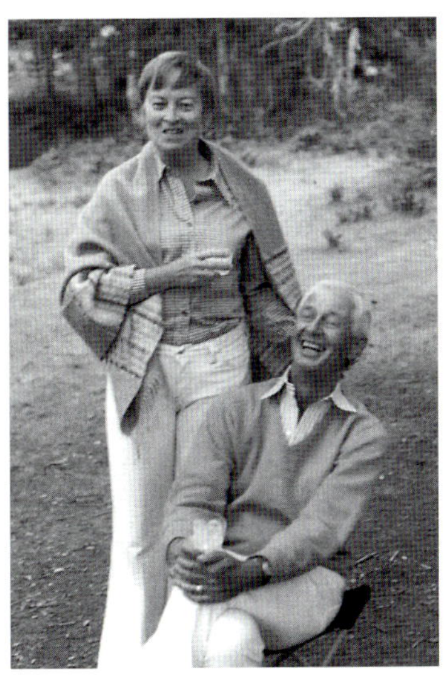

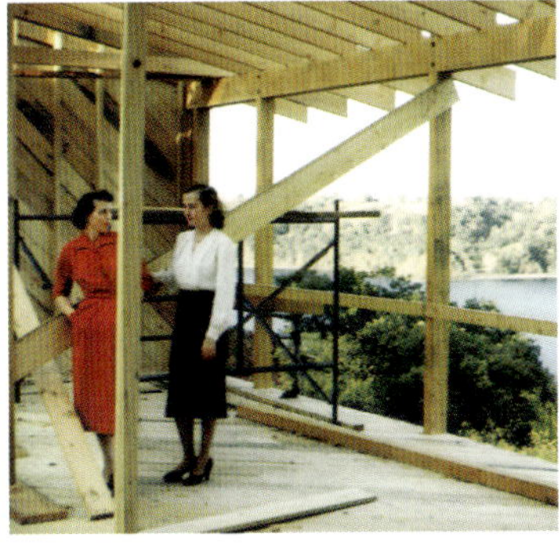

In her early years as a student at Wellesley College, Bliss planned to follow her model, Mme. Marie Curie. The impact of World War II descended, reports of casualties and the illness of her father began to shatter her world. Doubts emerged about the direction of her work, whether she should continue college or join military service. Her family, knowing her physical state and independent nature, delegated Julius Tolton, her favorite cousin and a captain in the air force, to dispel her illusions about the military.

With strong support from the faculty and medical staff, her energy and focus were restored. A period of exploration of new areas followed: writing, biblical history and art history along with mathematics. Her war work shifted from hospitals to summers working on drawings of warships for Gibbs and Cox, naval architects in New York City.

She met Roy Creighton, the architect father of her classmate, on his return from imprisonment in occupied China. He encouraged her to examine her range of interests and explore the possibilities in architecture. An interview with Dean Joseph Hudnut of Harvard's Graduate School of Design confirmed her eligibility. He offered the opportunity current with Harvard students to take her first year of graduate studies while a senior at Wellesley. After a strenuous year commuting between institutions, Bliss took off for New York City to get practical experience in an architectural office and enjoy museums and galleries. She also studied painting at the Art Students League and Statics at NYU School of Engineering to fulfill needed credits for Architecture.

Returning to Harvard she took advantage of a joint program with MIT to work with Gyorgy Kepes in the area of color and light. To help support her graduate studies, she became a "ghost writer" for a prominent Boston hospital architect doing his speeches and articles for publication.

Upon graduation she was given the assignment to do the colors and finishes for a huge VA hospital project. Concerned about the well being of the patients and her limited experience, she began consultations with Kepes and psychologists on the Harvard faculty. Research on the subject of color influence was very limited and inspired her future experimentation and writing on the subject. Her friendship with Anni and Josef Albers at this time was very influential. After completion of the project she took off for Europe with her husband, who had been awarded a Rotch Traveling Fellowship for architecture. A great adventure on bicycles followed.

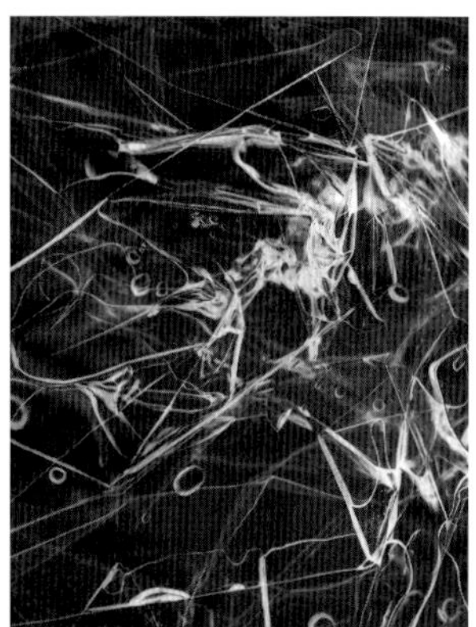

PHOTOGRAM 1947

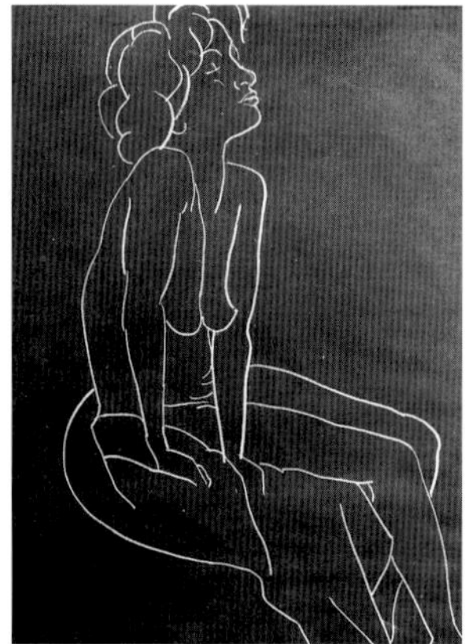

DRAWING 1954

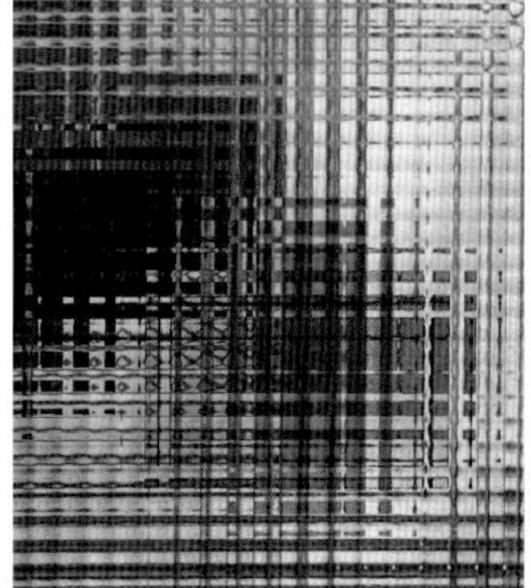

LIGHT STUDY 1947

Returning to Boston after stretching their resources to the limit, the Blisses resumed architectural work in different offices. Anna was called upon again to direct the color work for hospitals although anxious for a more rounded professional experience. Robert accepted an appointment to the faculty of Architecture at the University of Minnesota in 1952 and they established the firm of Bliss and Campbell, Architects there in 1956.

Involvement with the Walker Art Center and Minneapolis Institute of Art provided multiple stimuli for design work, enjoyment of art, and civic activity. Anna began to study painting in evening classes with artists Cameron Booth, Urban Couch and Bernard Arnest. Her short summer vacations from the office focused on printmaking at the University of Minnesota with Malcolm Meyers and Richard Haas.

With fellow members of the Walker's Arts Council they became involved with city planning and preservation issues. Highway expansion was threatening communities and open park space. Redevelopment was destroying large segments of the older section of the city without concern for historic and community value. Their office became the center for campaigns to save the 1893 Metropolitan Building and the 1908 Owatonna Bank of Louis Sullivan. They were defeated on the first due to strong pressures from commercial interests but rallied national support to preserve Sullivan's work.

In 1963 Robert was appointed Head of the Department of Architecture at the University of Utah , where he later established the Graduate School of Architecture as its first Dean. They arrived in Salt Lake City with barely enough time to settle in and to fly to Brazil. The Walker Art Center had been selected to organize the VII Bienal, Sao Paulo exhibition for the U.S. Information Agency and the Blisses to design the installation. Robert installed the complex works of the 10 sculptors: Agostini, Chryssa, Kipp, Schmidt, Decker, Sugarman, Mallary, Segal, Wines and Weinrib, while Anna supervised the installation of the paintings of Adolph Gottlieb, who won the international grand prize. He said it was his best ever exhibition. Before returning to SLC they made brief stops in Argentina and Peru for Macchu Pichu. The last inspired a return visit in 1969 on sabbatical.

Footloose in SLC after intense involvement in Minneapolis, Anna pursued language studies at the University of Utah, Anthropology with

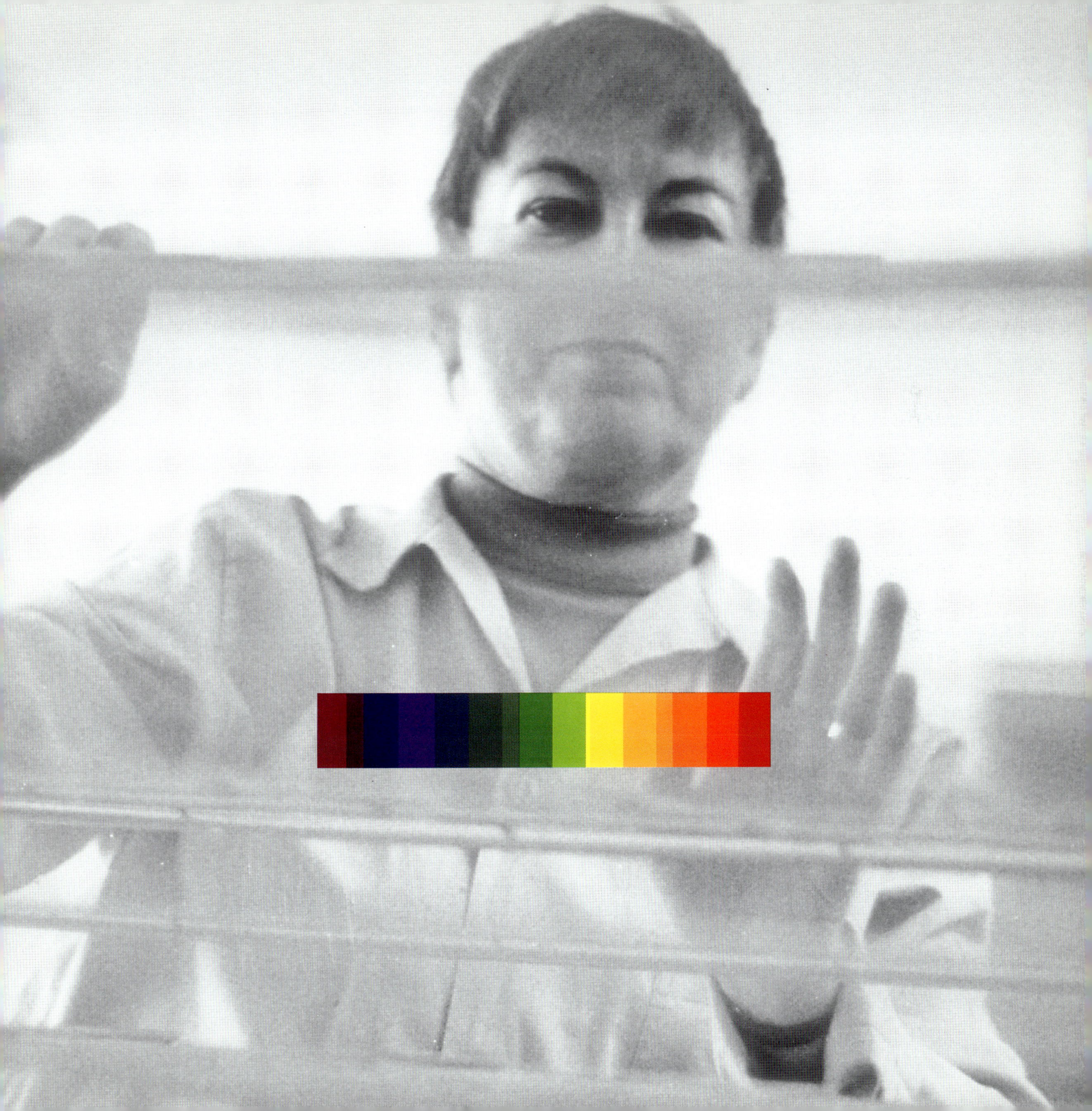

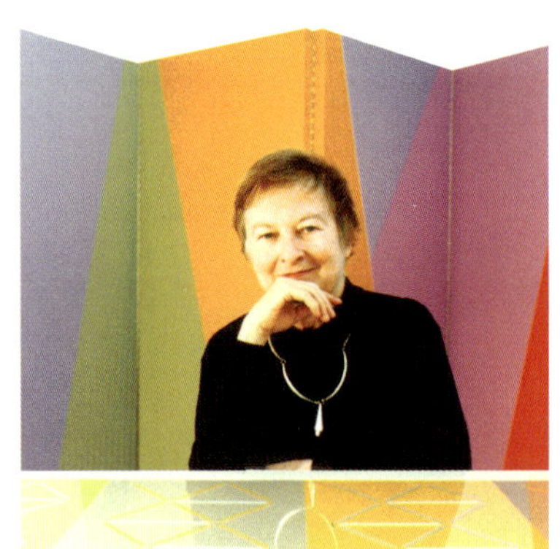

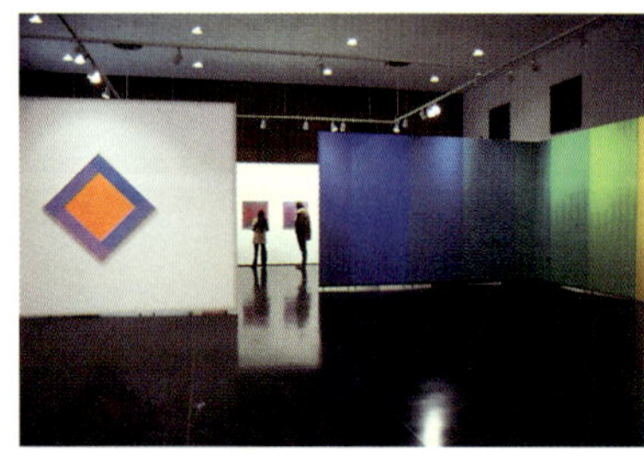

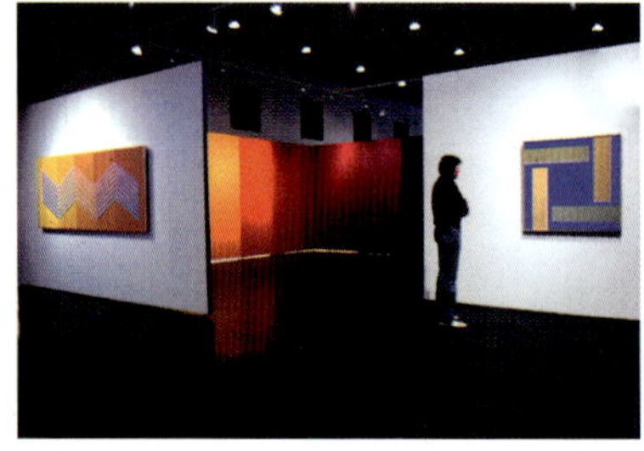

Charles Dibble, modern dance and computer studies. Her architectural work was limited locally to house remodelings for young professional people, prepared in stages to permit their financing over several years. She continued to work for Minnesota clients and a clinic she had designed also for various stages of expansion. The latter became a color lab as the doctors liked to have it refurbished every few years. She also lectured on color for a quarter as part of the reorganized curriculum in Basic Design for Architecture at the University of Utah.

While working on her important series: COLOR/LIGHT/MODULE, the president of NSID, an editor of Better Homes and Gardens came to photograph one of her small projects for the magazine. Seeing her color work, he asked her to represent the organization, later ASID on the Inter-Society Color Council. As a consultant, lecturer, and member of the Inter-Society Color Council, she traveled frequently from coast to coast. In 1975 she moved her art studio out of the basement of their house to a remodeled 1898 family grocery store. A surge of printmaking and large works on canvas followed.

Anna once told Josef Albers that she envied his squares – a simple form that permitted focusing completely on what color could do. By the time she had returned from the sabbatical in Peru, she had decided to concentrate on knowing color. While she admired Josef's exploration of color phenomena in his great Interaction of Color, his own work was largely devoted to the illusion of transparent layers of film color. She wanted to pursue areas that he had barely touched and to extend the experience of color into 3 dimensions for architecture.

For the publication of COLOR/LIGHT/ MODULE 1973 Bliss selected prints for two groups of studies that explore the range of color vibration usually associated with complementary hues of similar light value. She also distinguished a second area of vibration due to extremes of light intensity; ex. white & black or yellow and spectrum violet. Further examination was based on quantity: line widths and their repetitions, equal amounts of complementary colors in one portion of the print, one color dominating in another portion and the reverse in another. Varying the width of lines and the range of color intervals developed movements that began to suggest companion units for their resolution and so the modular organization of SPECTRUM SQUARED came into being. Modules take many forms in the art of Bliss, whether expressed or implied . The triangular articulation of Series II reflects the concept of the study most clearly and in combination becomes more three dimensional in form.

These two series formed the basis of the exhibition of Color/Light/Module, Three Dimensions of Color at the Lowe Art Gallery, Syracuse University in 1976. She was also invited to do her first site specific work in this instance a giant walk-in spectrum, CHROMATIC CONTINUUM, 120" h x 720"w. Many color illusions that can be seen at a small scale fail to develop at an architectural scale. She found the problem in pigment saturation. Here she used screen paint. In later large scale paintings she would use several coats of Winsor Newton oils to achieve density and color space. The spatial feeling of color, very important in her larger works, is something she shared with Adolph Gottlieb. He advised her to use rollers for applying the pigment to avoid surface textures that would destroy the illusion of space.

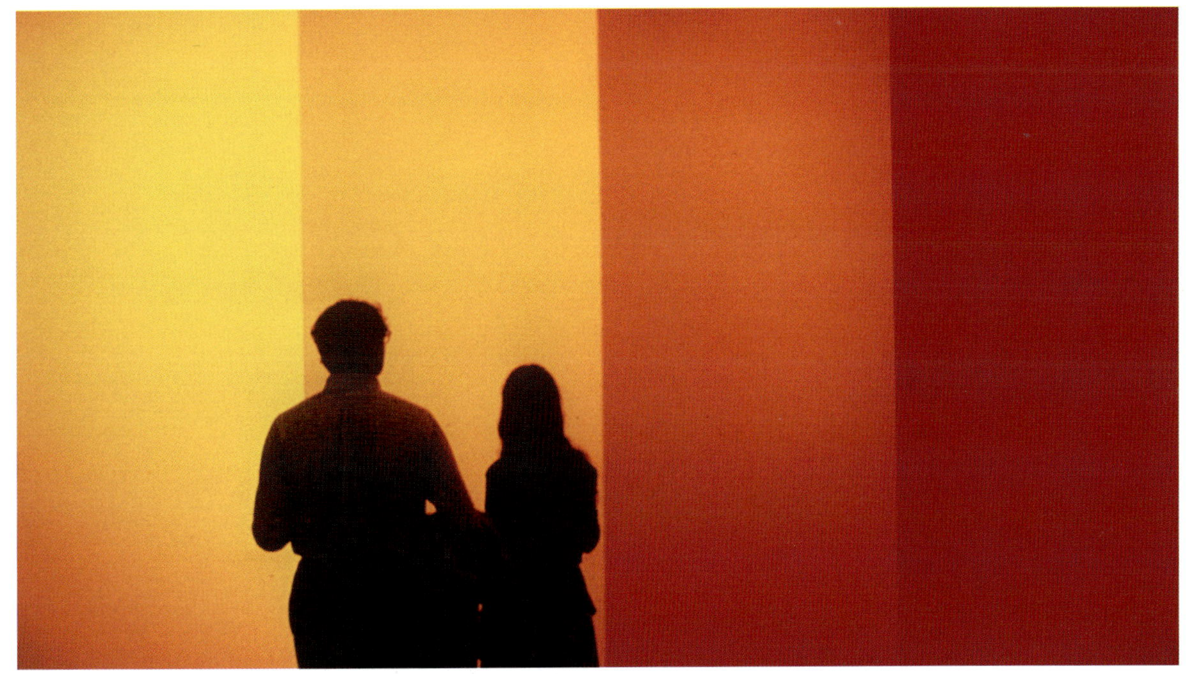

Color/Light/Module 1976
SCREENPRINT ON WOOD PANELS
120" X 720"
LOWE ART GALLERY EXHIBITION

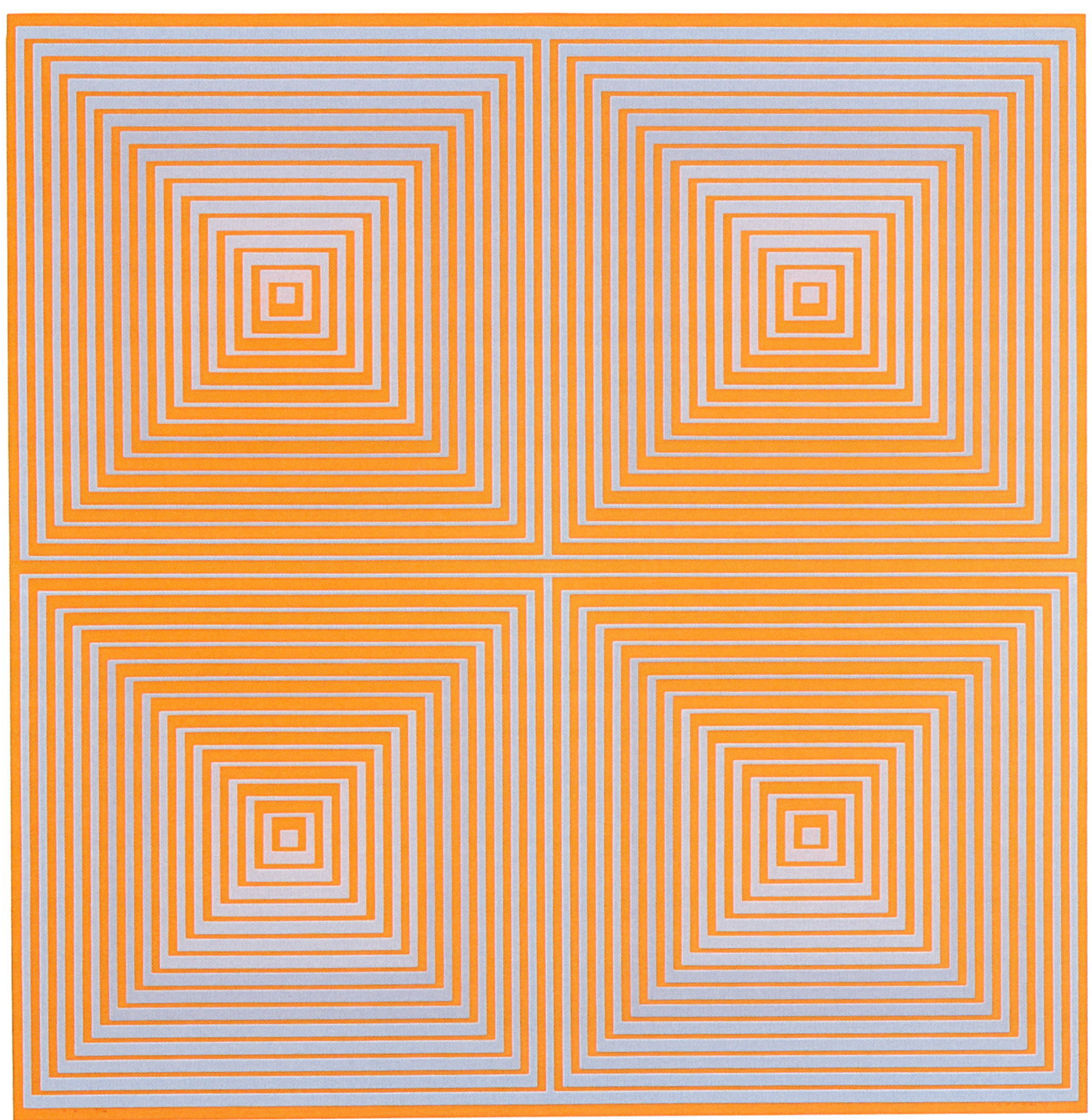

SPECTRUM SQUARED
1973 SCREENPRINT 31"x 31"
VAR. C, SERIES III

THE SERIES DEVELOPS A GRADUAL SEQUENCE OF THE ENTIRE SPECTRUM WITH MAJOR HUES DOMINATING. THE SIMPLE DESIGN STRUCTURE PERMITS THE DIRECT EXPERIENCE OF COLOR CHANGES.

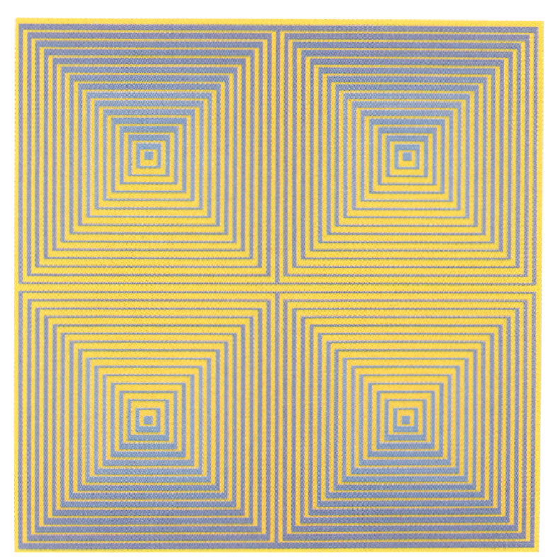

Variation A

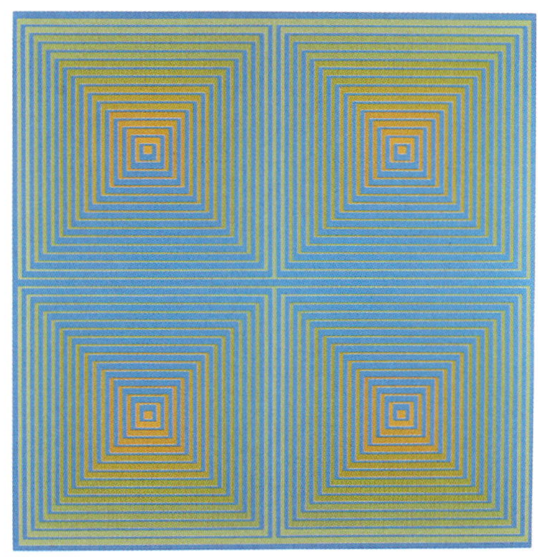

Variation J

17

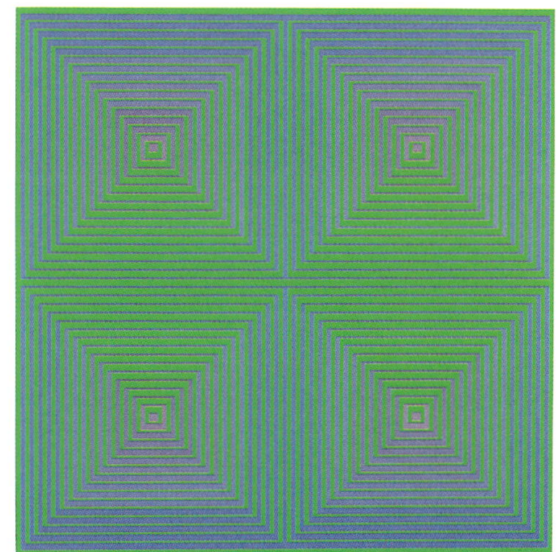

Variation L

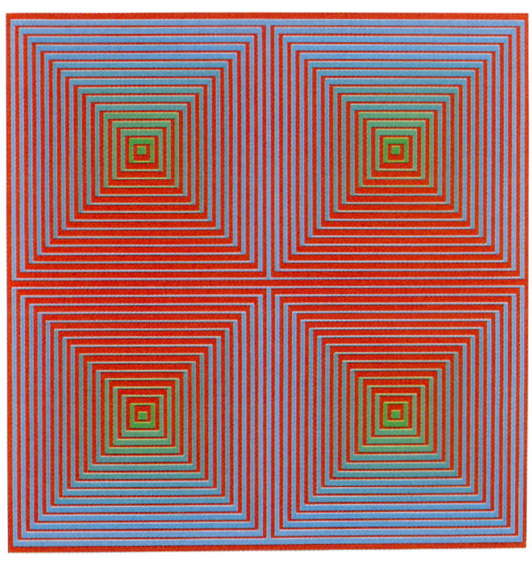

Variation E

**Modular Combination Series II
Triangular Articulation**

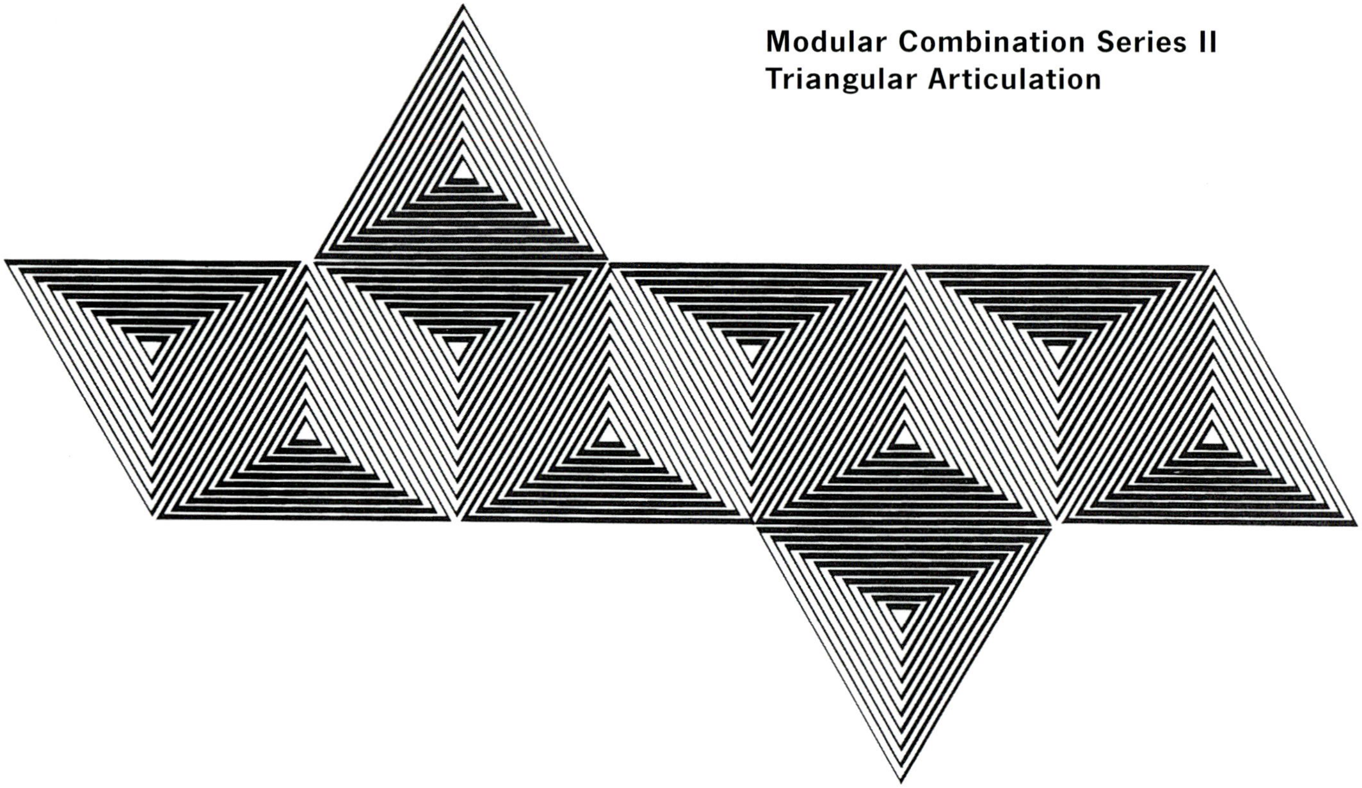

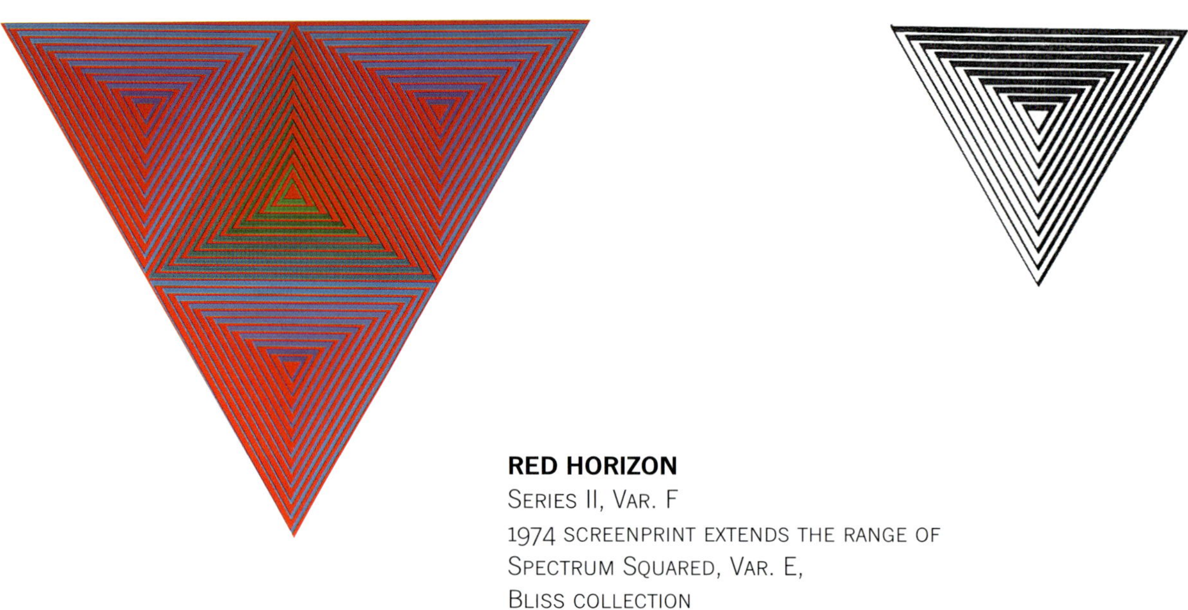

RED HORIZON
Series II, Var. F
1974 screenprint extends the range of
Spectrum Squared, Var. E,
Bliss collection

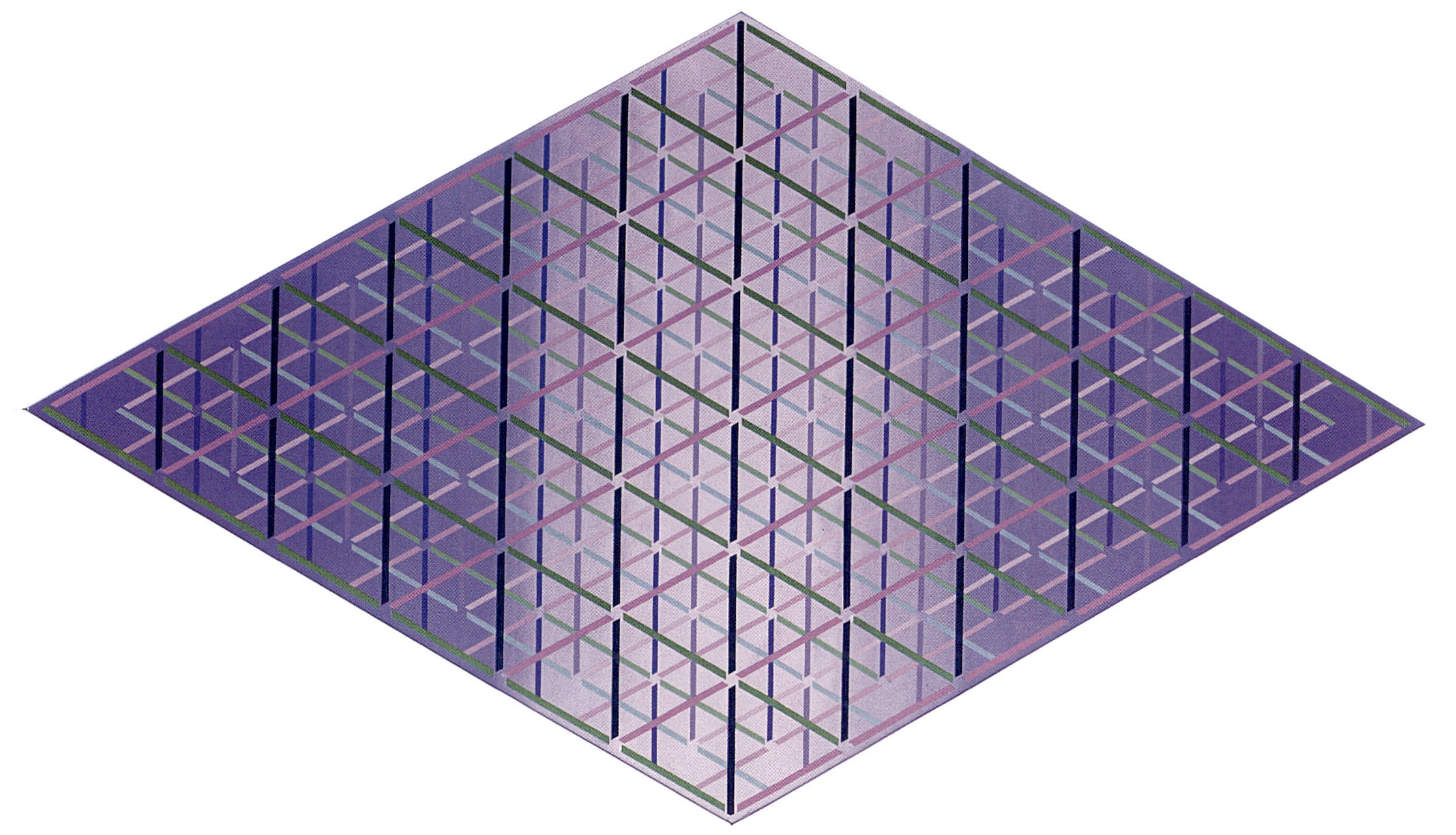

ECHOES III

1981-82, 40" x 70" OIL ON CANVAS BRINGS TO A RESOLUTION SEVERAL STUDIES DEVELOPED INITIALLY IN SCREENED PRINTS AS WELL AS OILS. AGAIN THE "THREADS" ARE MANY AS THE FORM IS REMINISCENT OF SERIES II ORGANIZATION WITH TRIANGULAR MODULES. CONCERNED ABOUT THE CLICHÉS THAT SURROUND COLOR AND THE ADVANCING AND RECEDING ASPECTS OF HUES, BLISS FOCUSED ON THEIR SPATIAL POSITION. HER STUDIO SPACE PERMITTED STUDYING WORKS AT DIFFERENT VIEWING DISTANCES. SHE FOUND THAT HUE IS NOT ALWAYS THE DECIDING FACTOR. OFTEN THE LIGHT-DARK INTENSITY AND SATURATION MAY CONTROL THE SPATIAL POSITION OF A COLOR WITHIN A PARTICULAR CONTEXT. ONE CAN COMPARE THIS WORK WITH THE SCREENPRINT, *CHAN CHAN*, WHERE THE FORM DISAPPEARS ON CLOSE VIEWING.
COLLECTION: MICHAEL ADAMSON

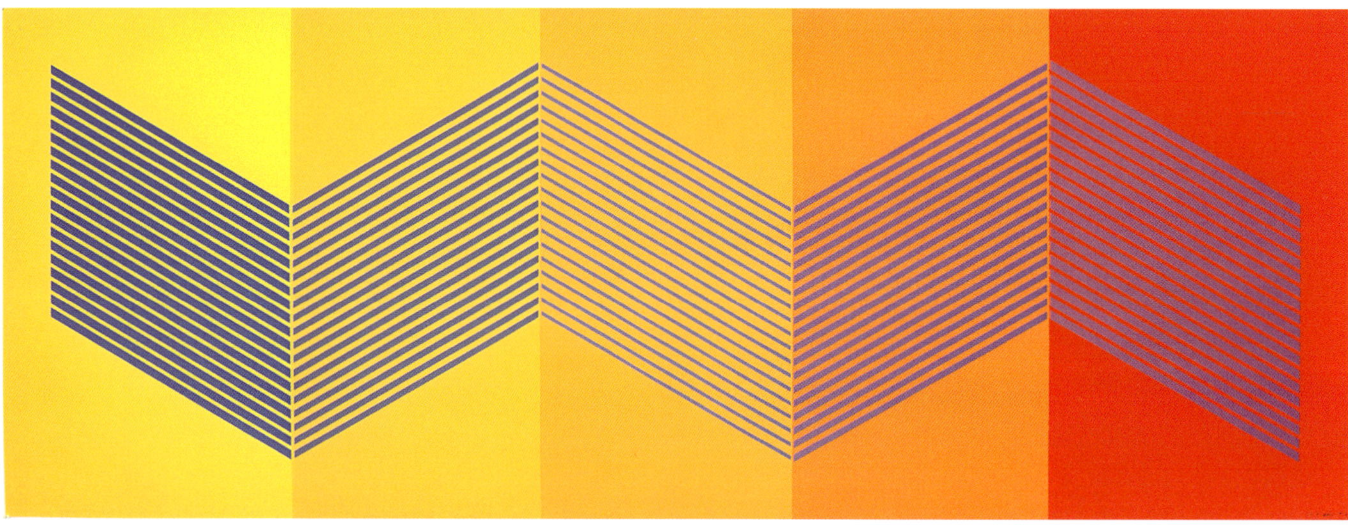

SOLAR SCREEN

1976 SCREENPRINT, 33" X 86"-
HERE SHE CREATES STRIPES OF THE FIGURE TO STUDY
COLOR CHANGES WHILE PRESERVING THE MOVEMENT.
COLLECTION: MICHAEL ADAMSON

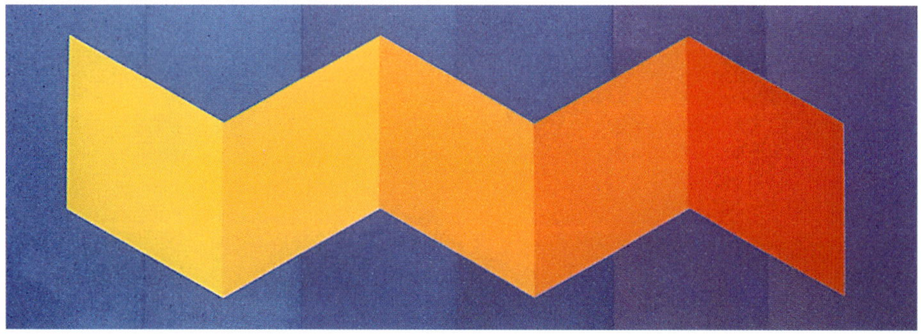

SOLAR DAY

1976 SCREENPRINT IS THE SIMPLER IN CONCEPT WHERE SHE EXPLORES
COLOR CHANGES OVER A LIMITED AREA WHEN JUXTAPOSED WITH RELATED
HUES. THE SOLID FIGURE (TOP IMAGE) IS OVERLAID ON A BACKGROUND
DEVELOPING A DIFFERENT RANGE OF COMPLEMENTARY COLORS.
COLLECTION: ROGER & JEEN BROWN

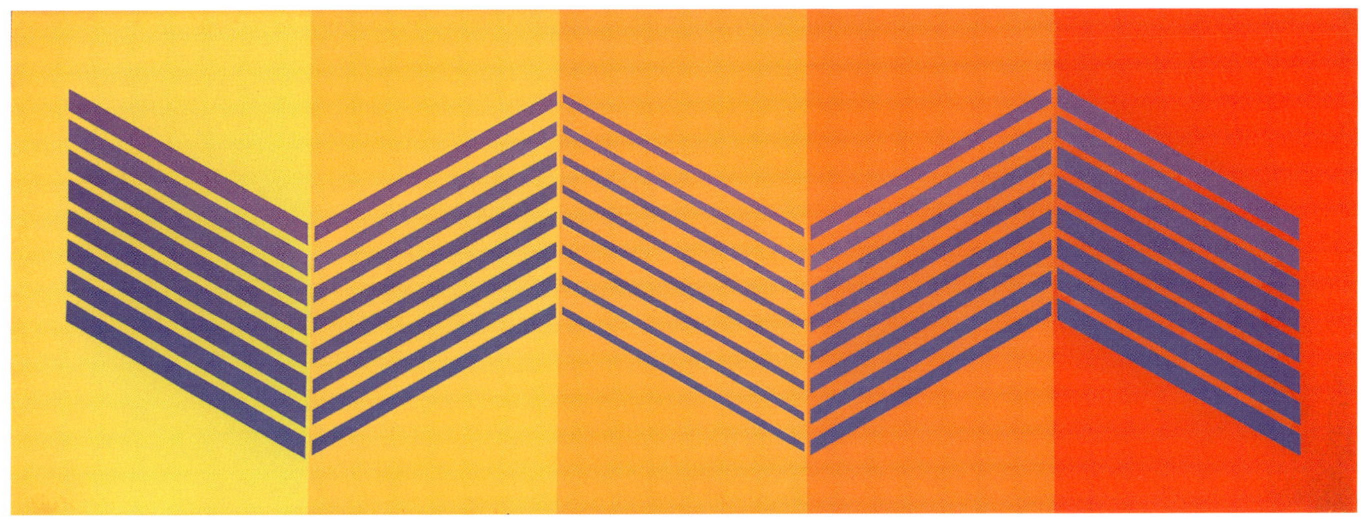

SOLAR BOUND

1976 OIL ON CANVAS, 43.5" X 113.5" IS THE MOST COMPLEX OF THE GROUP. THE STRIPES ARE CHANGING COLOR VERTICALLY AS WELL AS HORIZONTALLY. HER INTEREST IN THIS GROUP OF WORKS WAS SEEING AT WHAT POINT CERTAIN PHENOMENA TAKES PLACE AND HOW MUCH OPTICAL INTERFERENCE CAN BE INTRODUCED BEFORE THE ILLUSION IS DESTROYED. "SO THERE ARE MANY THREADS INTERWOVEN IN A WORK, BUT THE ART ITSELF MAY HAVE QUITE A DIFFERENT IMPACT THAN ONE'S EARLY INTENTIONS. I THINK OF THE **SOLAR** GROUP AS JOYOUS AND LIFE-AFFIRMING ".
COLLECTION: BRITISH PETROLEUM

Several works of varying nature have SOLAR as part of their titles coming from a fleeting reaction to her exhibition of prints in 1971 at the Salt Lake Art Center. Three of the early Series III prints had a sense of unity that suggested the rising and setting sun. The form of the alternating parallelograms, reversing itself in space came from Series II prints. She liked the movement that developed when combining prints.

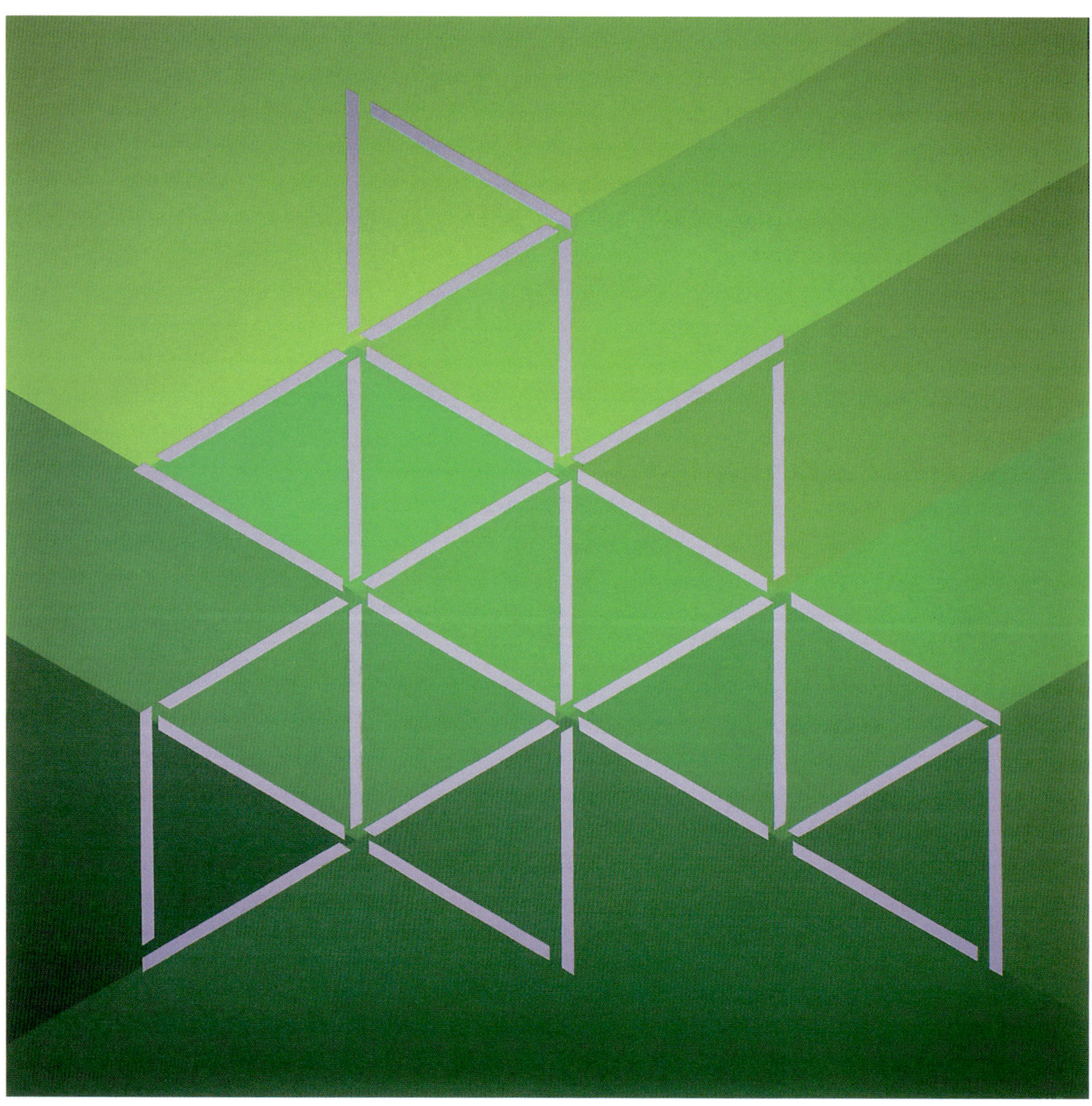

STRUCTURE IN GREEN SPACE
1977 OIL ON CANVAS, 80"x 80"
COLLECTION: SALT LAKE COUNTY

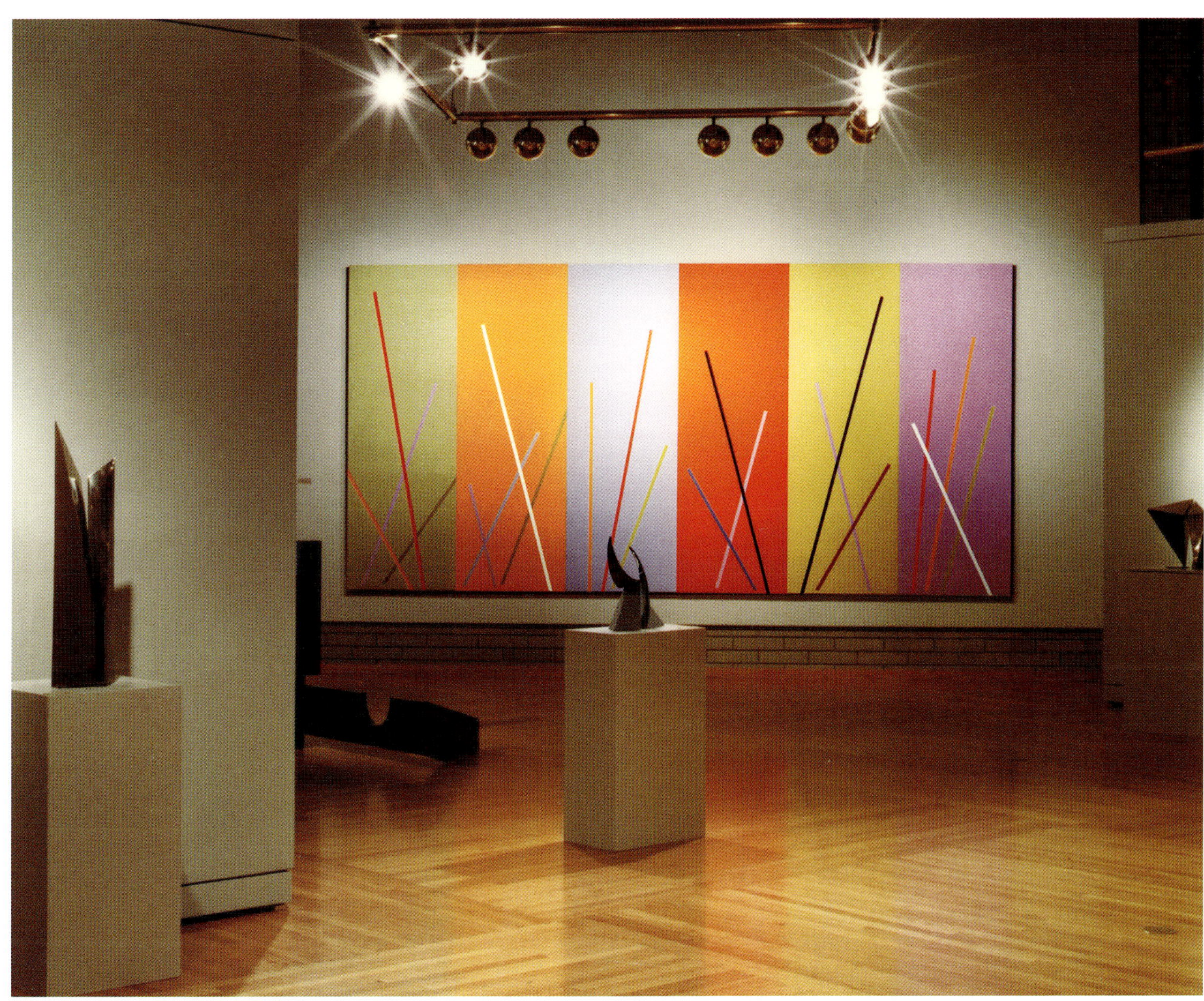

ORIENTAL SUITE

1982 OIL ON CANVAS, 8'-5" X 17', CREATED FOR A MAJOR EXHIBITION IN 1983 AT THE SALT LAKE ART CENTER.
BLISS IS VERY CONSCIOUS OF SCALE IN PLANNING HER EXHIBITIONS OR PERHAPS IT IS THE ARCHITECT EMERGING.
THE BEAUTIFUL PROPORTIONS OF THE SCREEN SUGGESTS EARLIER JAPANESE INTERESTS AND THE EFFECT OF
VIBRATING STRIPES GIVES ONE THE SENSE OF WALKING THROUGH FIELDS OF GIANT GRASSES.
COLLECTION: MATTHEW BOUSQUETTE

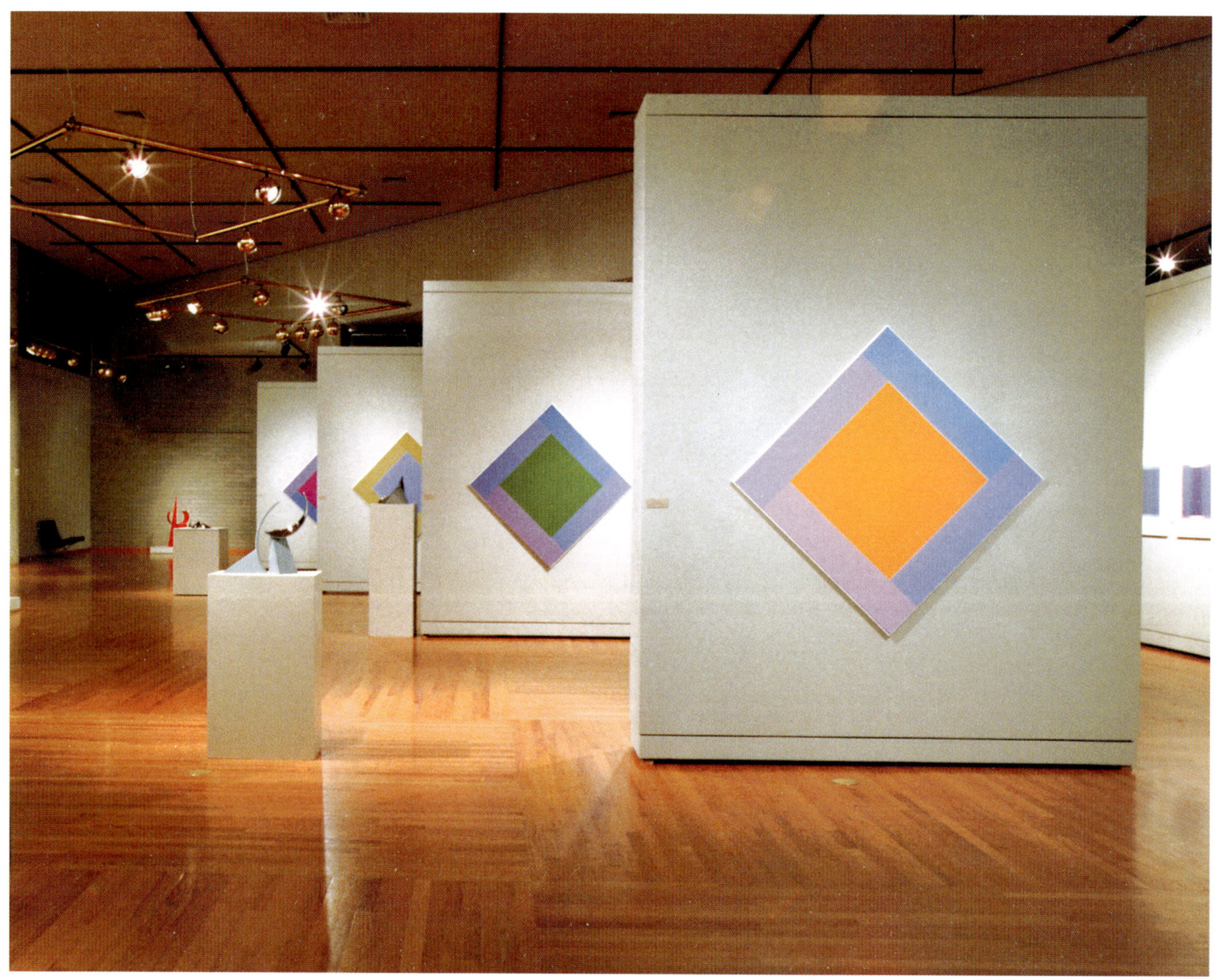

The PANE SERIES

1976-77, OIL ON CANVAS,
SALT LAKE ART CENTER, 59" X 59"
THE PAINTINGS DEPARTED FROM THE LINEAR DEVELOPMENT OF SERIES III PRINTS GIVING MORE EMPHASIS TO THE COLOR FIELD. THE FOUR OUTER RECTANGLES ARE WITHIN THE COMPLEMENTARY RANGE OF THE INNER SQUARE BUT THE VARIATION PRODUCES A TORQUE EFFECT. THE SPATIAL POSITION OF THE INNER SQUARE BECOMES AMBIGUOUS, AT TIMES ADVANCING, AT OTHER TIMES RECEDING OFTEN INFLUENCED BY THE SPECTRUM OF THE LIGHT SOURCE. AMBIGUITY IS A QUALITY OFTEN NURTURED IN HER WORK.

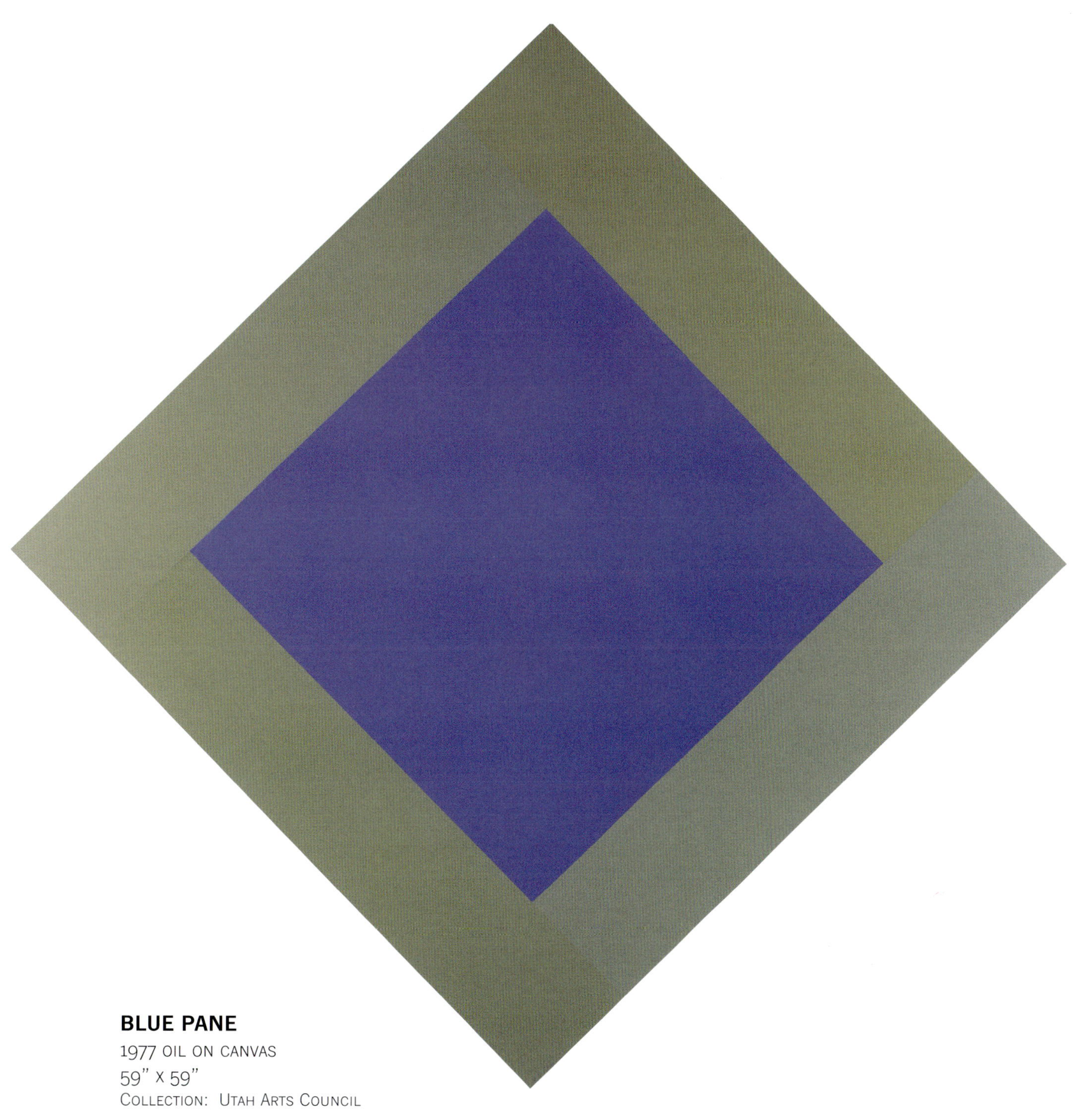

BLUE PANE
1977 OIL ON CANVAS
59" X 59"
COLLECTION: UTAH ARTS COUNCIL

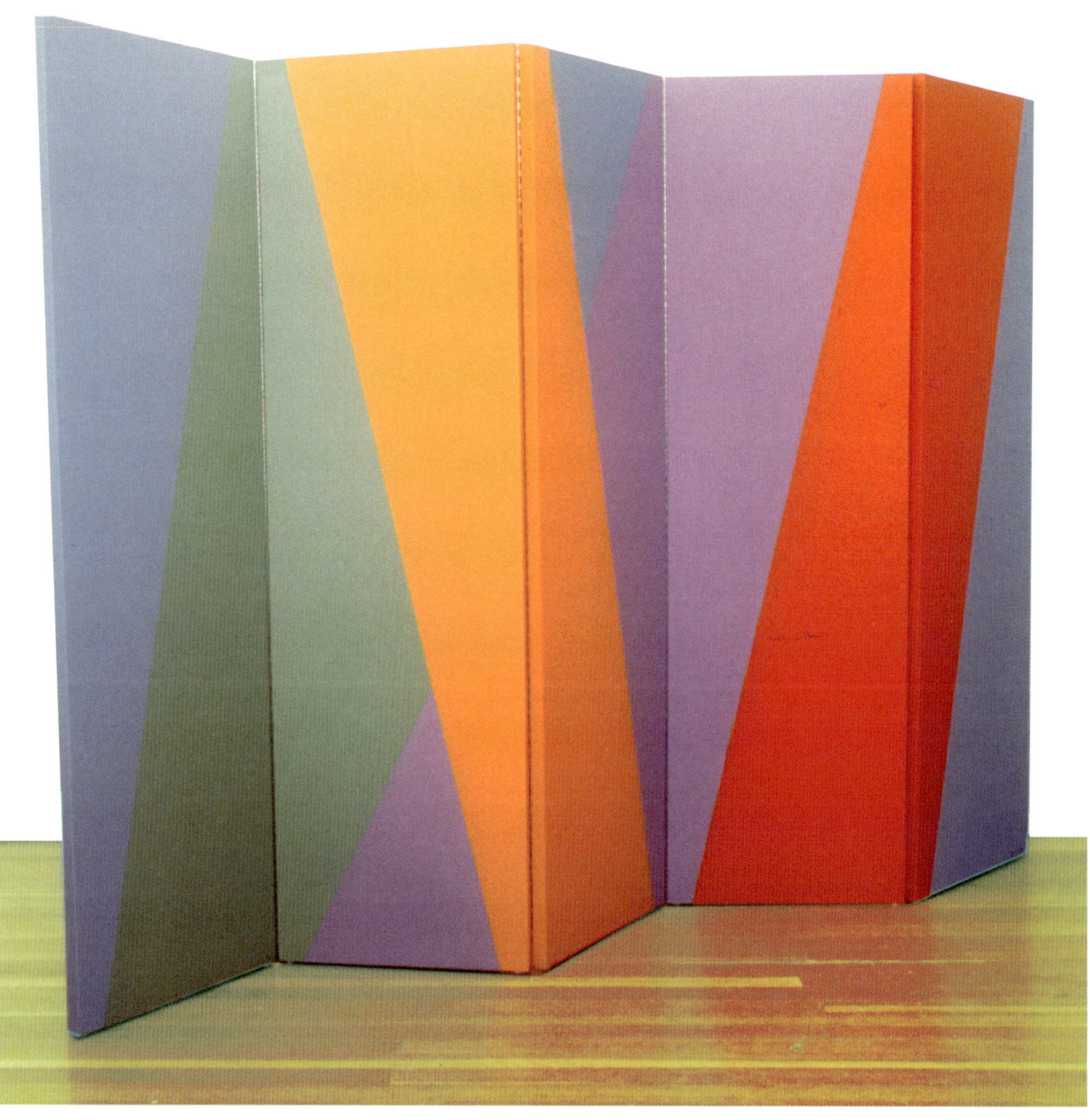

CELEBRATION

1985 OIL ON CANVAS FOLDING SCREEN, 6' X 8'- 6.5". A LATER WORK THAT REFERS
BACK TO THE COLOR FIELD STUDIES AND ADDS A GREATER EMPHASIS ON THE QUANTITY
OR AREA OF A PARTICULAR HUE. IT ALSO RESPONDS TO THE ARTIST'S DESIRE TO MOVE
ART FROM THE WALL TO BECOME A 3 DIMENSIONAL EXPERIENCE.
COLLECTION: UTAH MUSEUM OF FINE ARTS

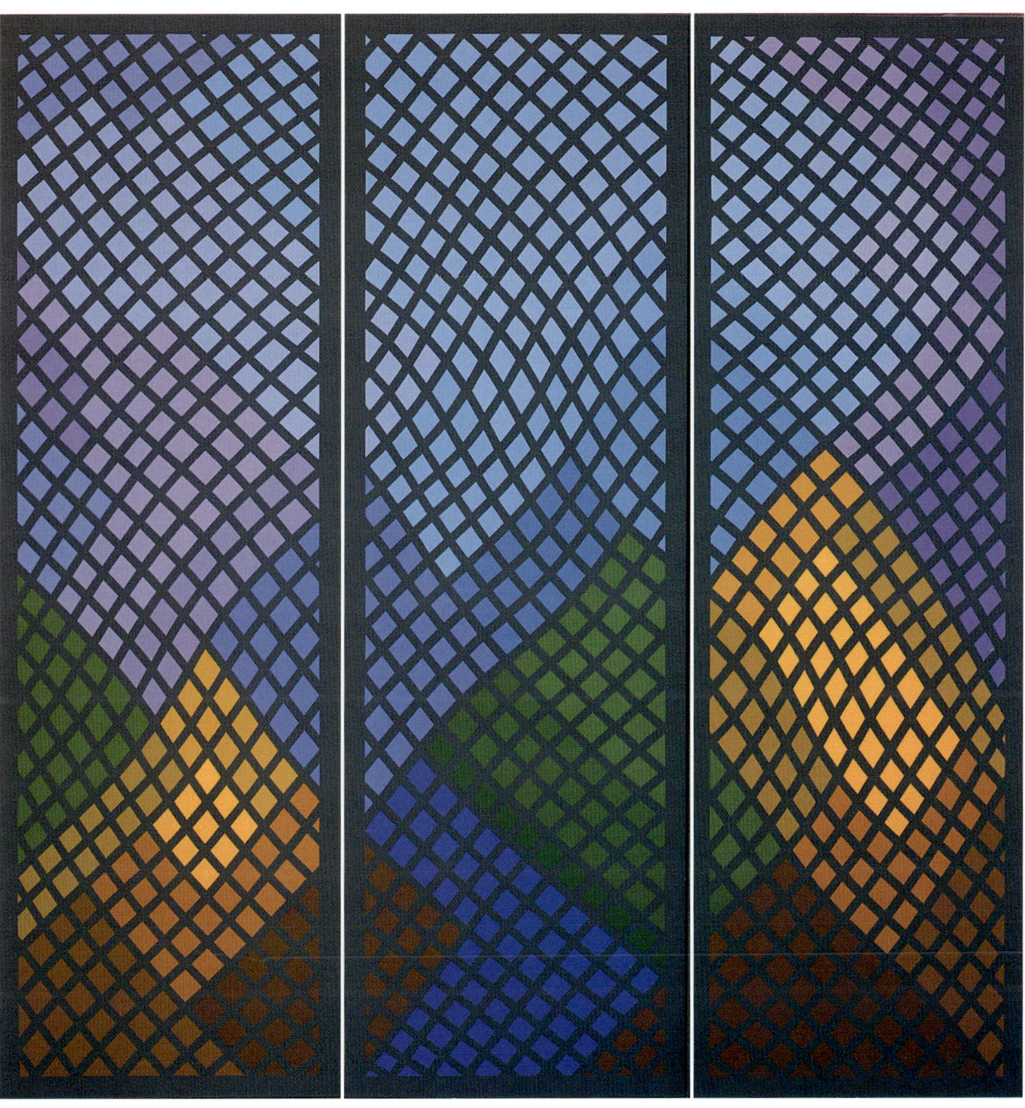

TRELLIS

1987 OIL ON CANVAS FOLDING SCREEN 7' X 7'- 1.5 ". "SITTING ON A SCREENED PORCH,
LOOKING THROUGH THE TRELLIS TO FLOWER GARDENS, SQUINTING YOUR EYES TO SEE
MERGING PATTERNS. FORMS APPEAR BEHIND THE OPEN SCREEN AND SOMETIMES READ
ON THE SURFACE. ANOTHER LOOK AT SPATIAL POSITION OF COLOR.
COLLECTION: DONALD & JANE STROMQUIST

**STUDY FOR THE
LAST BILLBOARD**
1974 SCREENPRINT
10" X 22 1/2"
EDITION. MUSEUM WITHOUT
WALLS INVITATIONAL
BILLBOARD SIZE.

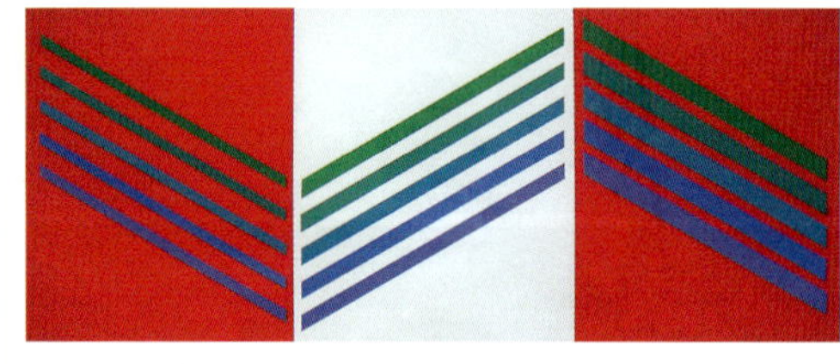

Collection: Peter & Hon. Judy Atherton

SOLAR DAY
1976-79 SCREENPRINT
11 COLORS
14 1/2" X 40"

LUNAR
1976-79 SCREENPRINT
14 1/2"X 39 5/8" IMAGE

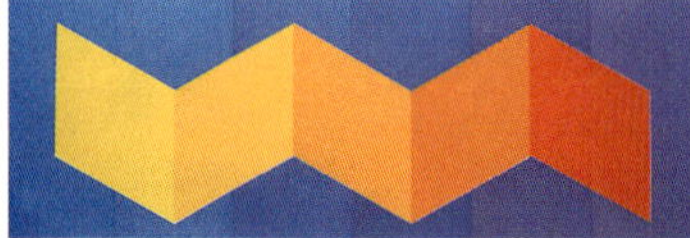

Collection: Roger & Jeen Brown

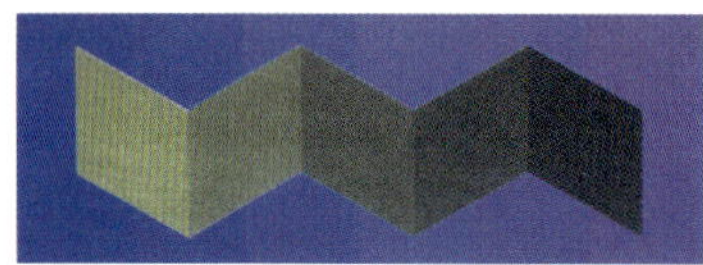

Collection: Roger & Jeen Brown

CORN ROW
1982 SCREENPRINT
32" X 50" IMAGE
(FRAMED 40"X 60")
EDITION

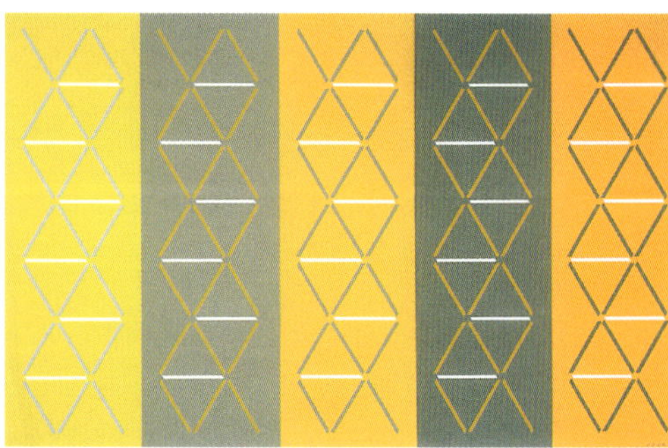

Bliss Collection

PRINT GALLERY:

For the artist screenprinting is a medium for exploration and
color research. Many of her ideas have been developed further
in the paintings. Finding limitations in the process she began
to experiment with the computer for ease of visualization.

SPIDERWALK

1983 SCREENPRINT
38 1/2" X 25 1/2"
EDITION, COMPUTER
GENERATED SCREEN

MIRAGE

1985 SCREENPRINT & AIRBRUSH
22" X 30" IMAGE
(FRAMED 24"X 32")
COMPUTER GENERATED SCREEN

CHAN CHAN

1981 SCREENPRINT ON PAPER
14" X 23 1/2" IMAGE
(FRAMED 21 7/8"X 31 1/4")
PRE-COMPUTER

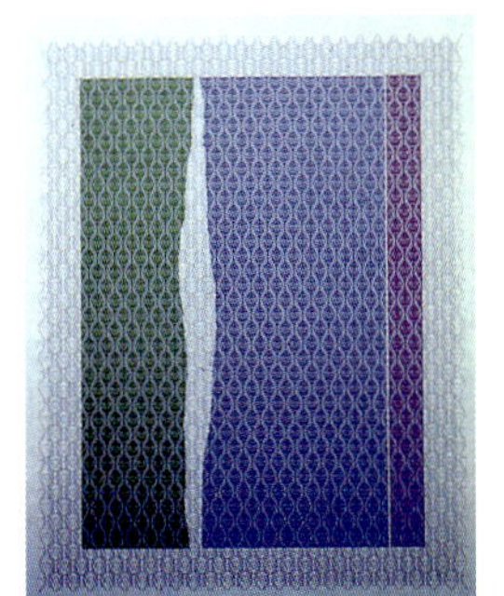

Collection: Noel & Klancy DeNevers

Bliss Collection

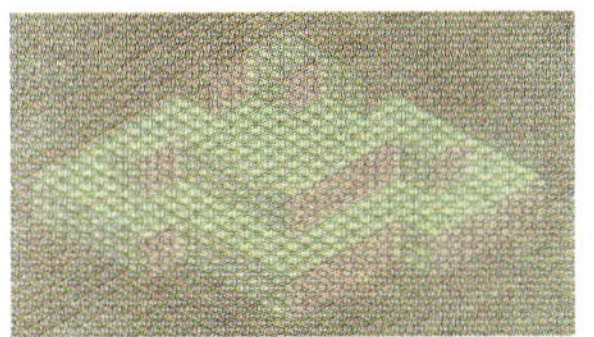

Collection: John R. & Sue Smith

OTHER WORLDS II

1993 EXPERIMENTAL
MONOPRINT ON RAG PAPER
22"X 11"

NIGHTSHADE III

1993 EXPERIMENTAL
MONOPRINT ON RAG PAPER
22" X 11"

VIEWPORT

1992 EXPERIMENTAL
MONOPRINT ON RAG PAPER
13.8"X 21.8" IMAGE
(FRAMED 20 1/2"X 28")

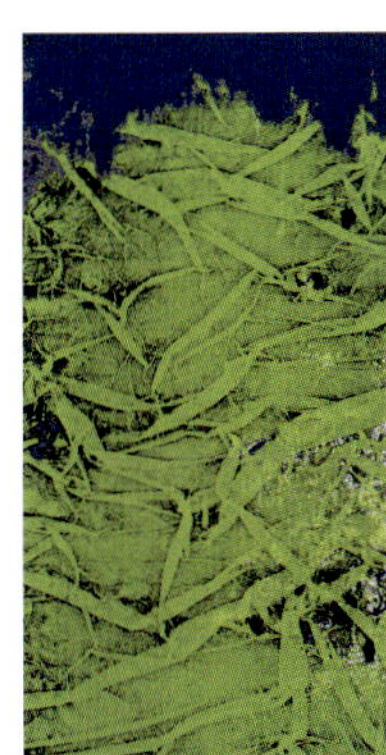

Bliss Collection

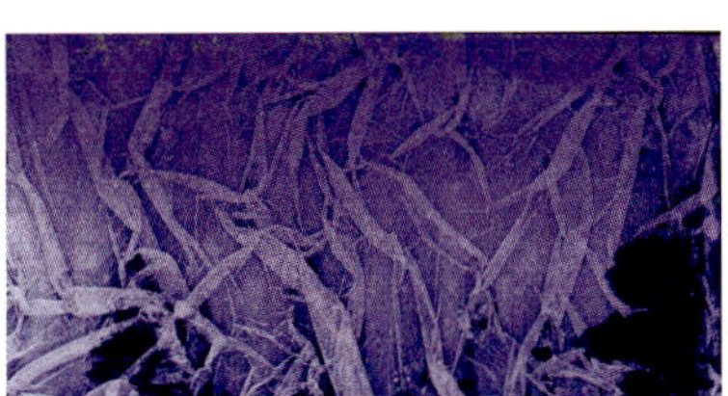

Bliss Collection

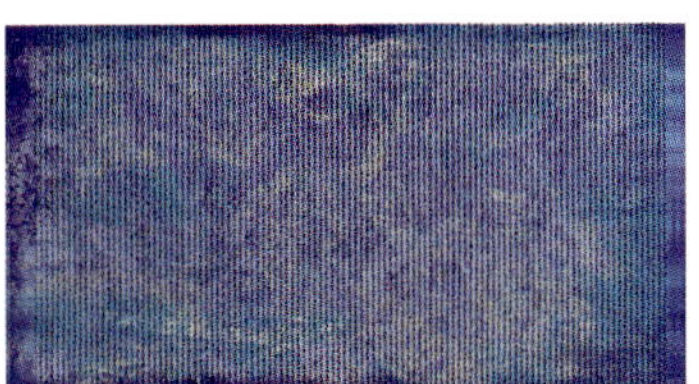

Collection: Utah Museum of Fine Arts

EDO TIGER AWAKENS IN FRACTAL FOREST

1993 SCREENPRINT
21" X 25" IMAGE
EDITION, COMPUTER
GENERATED SCREEN

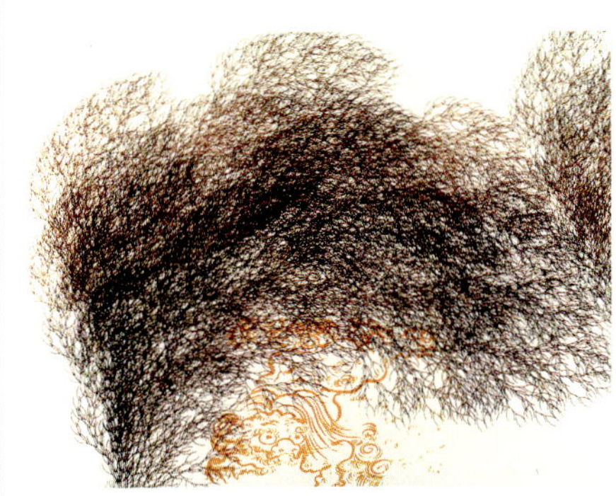

Collection: Edgar John Davies

JUMPING KOI AT LAKE TRUCHET

1993 SCREENPRINT &
AIRBRUSH
23 1/2" X 32" IMAGE
EDITION, COMPUTER
GENERATED SCREEN

CASCADE

1992 EXPERIMENTAL
MONOPRINT ON RAG
PAPER, 23"X 13 1/2"
(FRAMED 31"X 19")

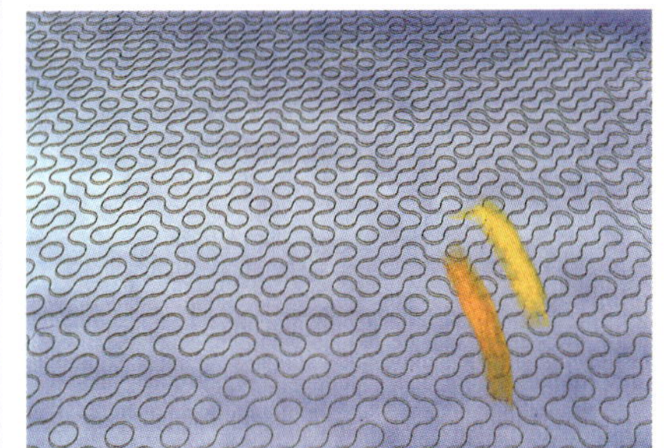

Collection: Utah Museum of Fine Art

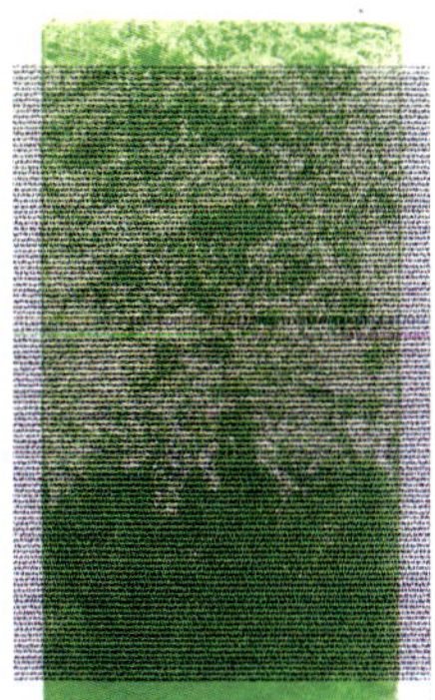

Bliss Collection

PASSAGES 1978-80

An Experience in Time, Color and Space

Rushing

TO ONE'S DESTINATION: A MOMENT OF SERENITY
TO CATCH A BREATHE.

Leaving

A JOYFUL INTRODUCTION TO THE CITY.

Moving

FROM ENTRY TO LOBBIES TO LANDING: CONTINUOUS
VISUAL INTEREST AS TRANSLUCENT PLANES OF COLOR
OVERLAP ONE ANOTHER AND THE ARCHITECTURAL SETTING.

Changing

WITH SUBTLE MOVEMENT AS DOORS OPEN AND CLOSE,
PRODUCING AIR CURRENTS.

Watching

AS PEOPLE MOVE THROUGH THE TRANSLUCENT PASSAGES.

Creating

NEW DIMENSIONS.

PASSAGES OFFERS THE STILLPOINT AT WHICH ONE IS MOVING
THROUGH SPACE AND HAS A MOMENT TO SAVOR THE EXPERIENCE AND
THE SURROUNDINGS. A PROPOSAL FOR THE SALT LAKE CITY
INTERNATIONAL AIRPORT, STILL IN THE WINGS, WAITING TO FLY.

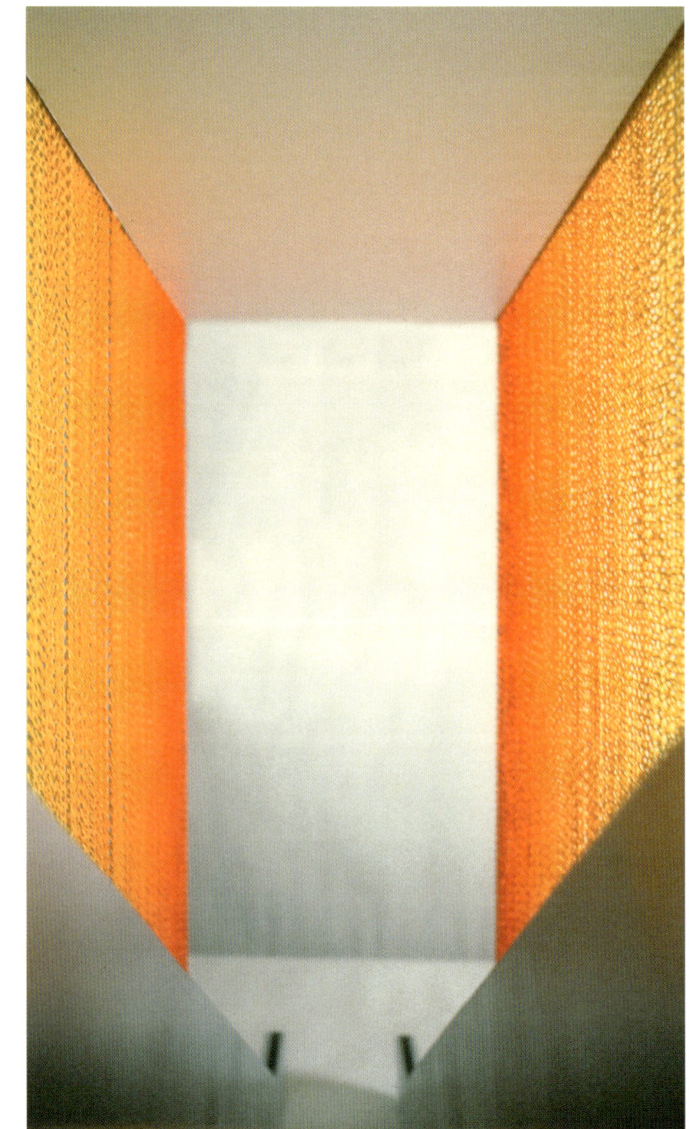

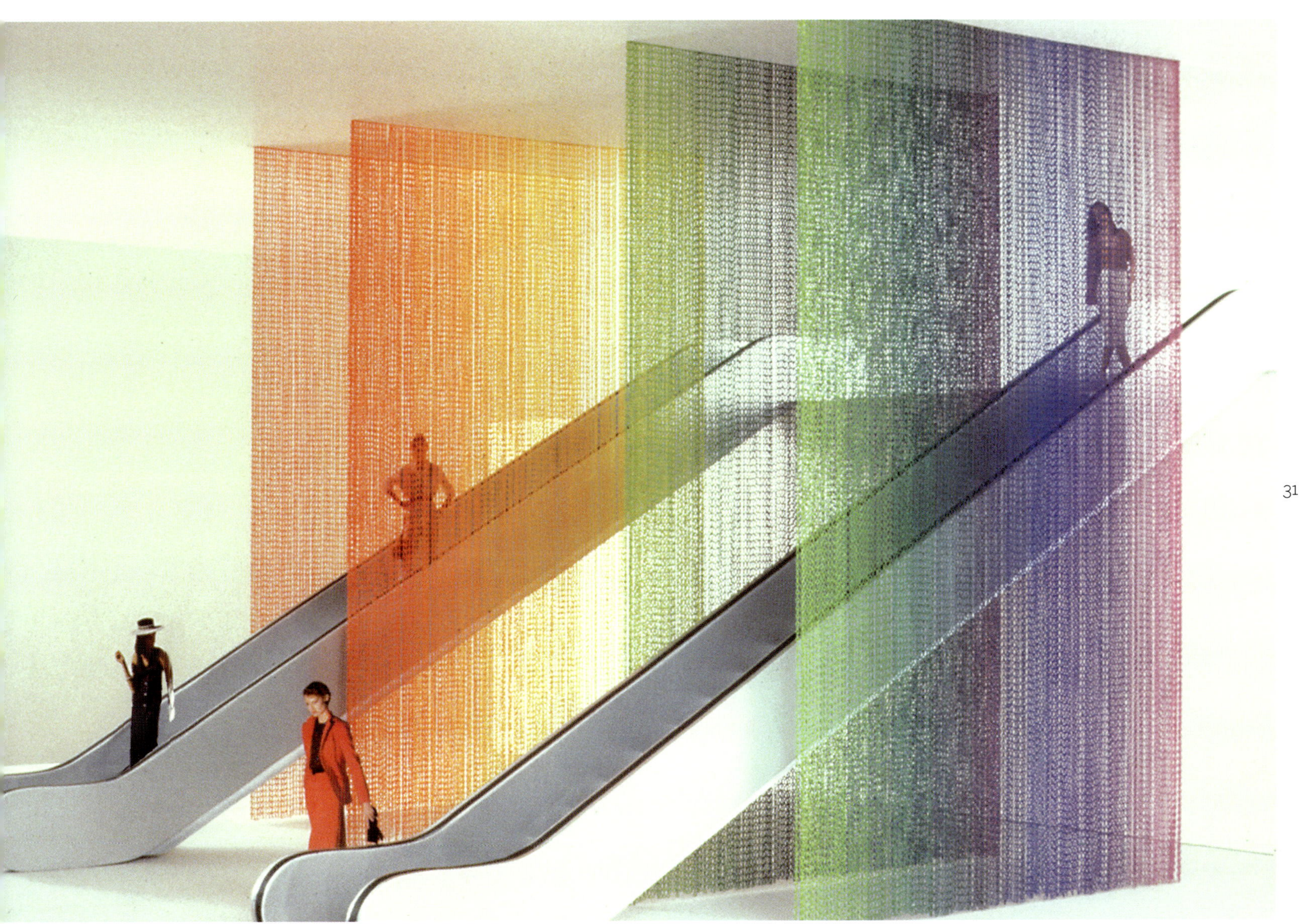

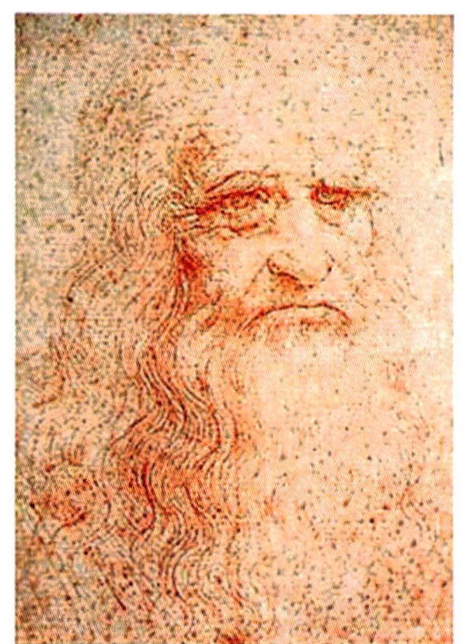

An intermezzo in Rome became a major stimulus that helped to define the future of Anna's art. Her earlier visit as a student in Europe had left her with a strong desire to return and become immersed in the rich environment of Rome. She applied and received a mid-career fellowship to the American Academy in Rome in 1984. Her program was to experience color and art in Renaissance Architecture.

"Those months in Rome literally saved my life," she said. "The change from a western city in a desert to an ancient river civilization, the light in the courtyard of the Villa Giulia, the riches of the American Academy Library and the Vatican, the layers of history that one walks through daily in this city." Not least was the stimulating company of fellows on expeditions, the discussions over dinner and the marvelous food in simple restaurants. She often traveled alone or with a friend to Florence, Milan, Venice and the hill towns enjoying the freedom to move at will. An introduction to Count Panza di Biumo gave her the opportunity to experience his great modern collection installed in an 18th century villa outside of Milan.

Her focus on modern and early Christian art, along with African and Oriental interests, moderated her views of Renaissance artists. The experience of their art as part of the architecture in Rome restored her perspective. She became absorbed in the ideas of the 15th and 16th century artists, who were explorers of the visual world. They built on their studies of classical theorists and intense observation of nature. The desire to know and understand permeates their writings and notebooks and is resolved in their art. She was particularly drawn to Piero della Francesca and Leonardo da Vinci. She visited the town of Vinci to experience an extensive exhibition of models of his scientific concepts and Arezzo for Piero's great art. The knowledge of mathematics, so disdained by many contemporary artists, was an integral part of their vision.

Some of her favorite churches and palaces were visited many times. The great ceiling fresco in the nave of S. Ignazio by Pozzi and his trompe d'oeil dome were astonishing feats of color and illusion, that challenged her modern tastes. The churches of Borromini were wonderful geometric constructions that offered relief from an excess of surface decoration. The gardens of Villa d'Este and other villas in or near Rome also provided enrichment. Some unexpected discoveries were the Oriental Museum with its collection of Islamic Art and the Library of the Japanese Embassy. Her Etruscan studies brought her to the Villa Giulia, a favorite retreat after Sunday mornings visiting several churches and lunch in the Piazza del Popolo.

Anna Campbell Bliss returned from Rome invigorated by the experience and with renewed understanding of the bond between the arts and sciences. Her early studies with Gyorgy Kepes had laid the foundation. She now had renewed confidence to continue pursuing her own ideas and in her own way. Buoyed by her support from the visual explorers of the Renaissance she entered national competitions for projects in Salt Lake City. She also organized a major conference, "New Technologies of Art -

Where Art and Science Meet" for the Contemporary Arts Group of which she was president.

All her major works after 1984 were large site-specific commissions. All drew upon personal resources and references to other ages and cultures as well as contemporary technologies. Although commissions are random and unrelated, gradually a continuity develops generated by the values the artist is developing in examining issues and solutions that absorb her.

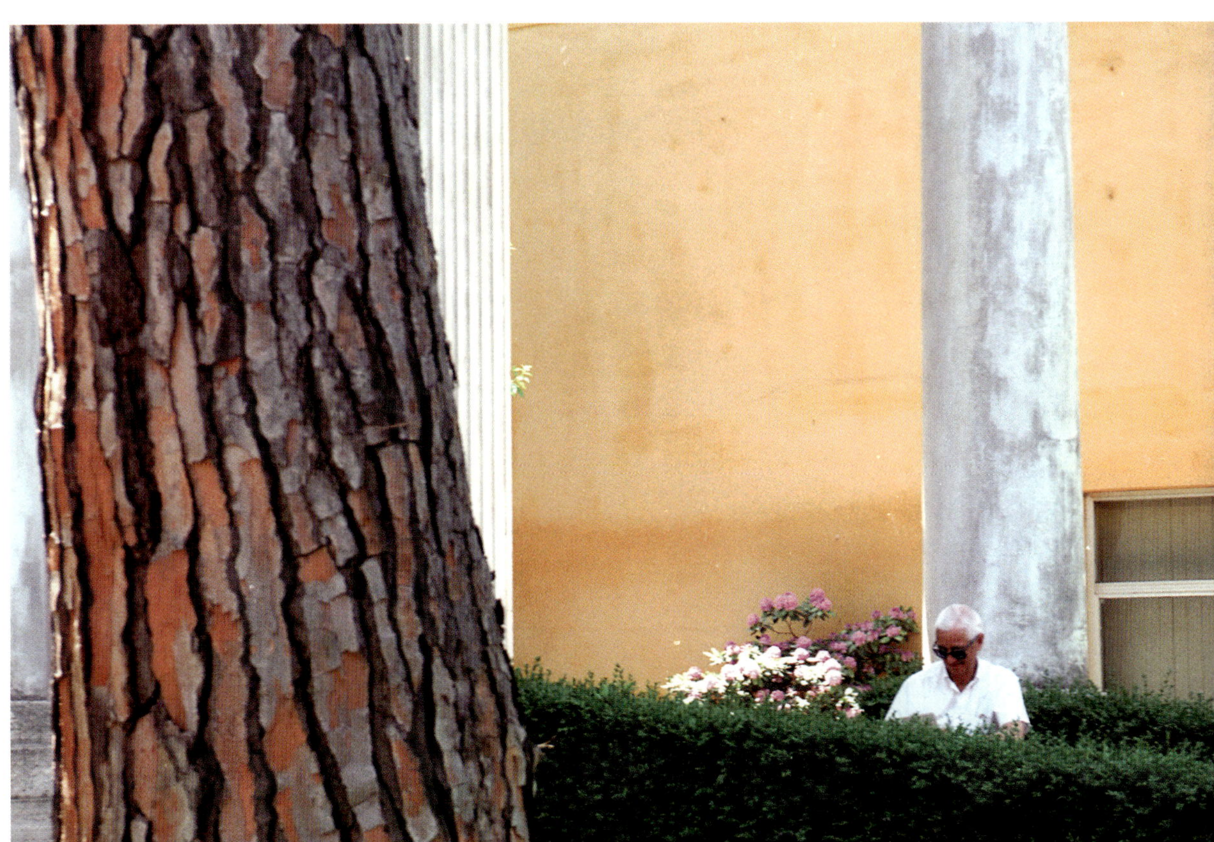

THE GARDEN OF
THE VILLA GIULIA

MAJOR SITE WORK

WINDOWS 1989 – 90 for the Utah State Capitol was her opening, using the computer as subject and major artistic tool to create the mural. Before installation the work was exhibited at the headquarters of the American Association for the Advancement of Science in Washington, DC. There it was nominated and became a finalist for a Computer World Smithsonian Award.

This was followed by LIGHT OF GRACE 1993, a stained glass window wall for St.Thomas More Catholic Church, another project receiving three AIA awards. DISCOVERERS 1996, a competition design for the Delta Terminal at the Salt Lake International Airport followed. Her most recent commission is her most extensive, EXTENDED VISION 2001-3, a mural that extends through 3 floors of the Cowles Mathematics Building at the University of Utah to create a "House of Mathematics". A catalog of the art is scheduled for publication in 2004.

From the Utah State Capitol to the International Airport, from St.Thomas More to the University, Anna has left her mark on the urban scene in which she lives. Like artist explorers before her she looks back to the traditions that nurtured her and looks forward to new challenges and the unknown. And like her predecessors she is multi-lingual speaking and writing through many disciplines to audiences around the world. (See her Seminars & Lectures in Chronology) With the completion of her commissions she has written for the magazine LEONARDO about the various ways in which art, science and technology meet in her work.

GREAT COLOR ARC

1995 EXPLORES COLOR AND LIGHT WITH CAST STAINED GLASS IN AN ALUMINUM FRAME, A PROPOSAL FOR THE SALT LAKE CITY INTERNATIONAL AIRPORT COMPETITION CREATING AN HEXAGONAL HONEYCOMB ARC, 9'-10" X 37' X 8'. LIKE PASSAGES 1978, THE GREAT COLOR ARC REMAINS ONLY AN IDEA IN MODEL FORM AND HAS NOT FOUND A HOME IN SALT LAKE CITY.

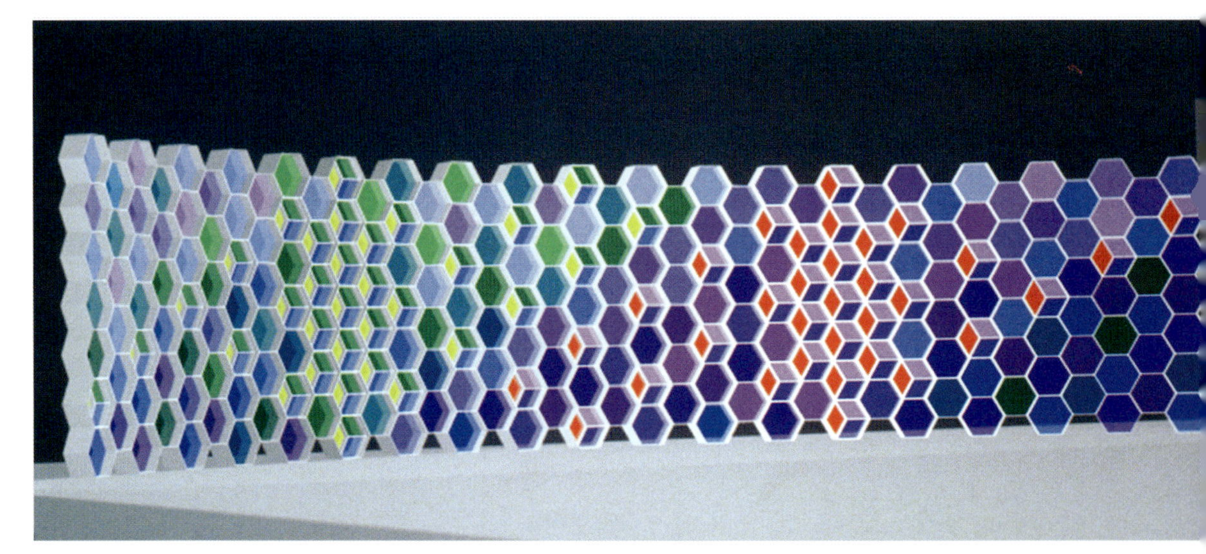

WINDOWS

1989 - 98
MIXED MEDIA,
ALKYD ON
ENAMELED STEEL

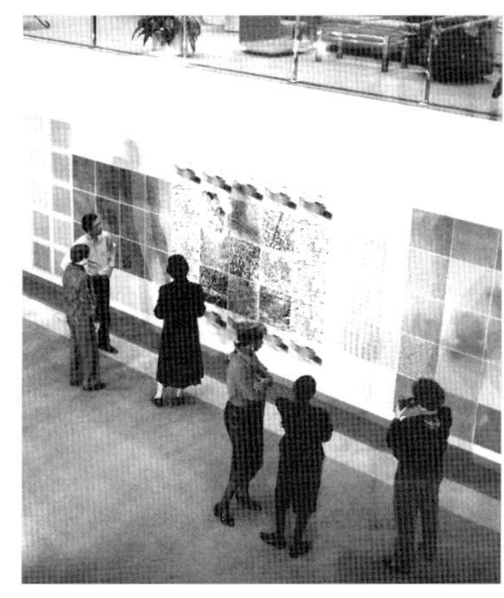

PHOTO OF INSTALLATION AT
THE AMERICAN ASSOCIATION
FOR THE ADVANCEMENT OF
SCIENCE, WASHINGTON DC,
PRIOR TO SALT LAKE CITY.

I am back on solid ground at the Utah State Capitol between government offices and the Data Processing Center for my third visit to WINDOWS. Having read the LEONARDO article, "Mathematics for the Garden of the Mind", I am beginning to appreciate the multiple perspectives within Anna's carefully composed garden. Her words from the article become my guide:

"Viewing distance, permanence and the broad range of imagery I was asked to explore led to my creating a multilayered, visual collage spanning time and place, drawing upon the past, the present and ideas for the future. A further organization of diverse images suggested the format of a giant computer screen with six windows: **Palette**, **Outerspace**, **Human Scale**, **Communication**, **Microworld** and **Fractalscape**. The final key in the execution of the mural was the modular unit...(enameled metal plates 18 inches square) which made it feasible to increase or decrease the viewing distance and size of images, similar to the "zoom" feature of a computer, and to combine them within one window." It also made screening feasible in combination with direct painting and greater control of color quality.

WINDOWS is a unique effort to bring the computer into the mainstream of art for a major mural at a prominent site. The computer was the inspiration, the subject matter and the major tool in creating the art. Textures and patterns with a mathematical base were programmed in the "C" language for plotting and transfer photographically to serigraph screens. Images were also processed and painted directly. To capture some of the brilliance of the computer screen, colors were mixed optically in a somewhat impressionistic manner by overlays of printing from the computer generated screens. (See **Palette**, Window #1) Specific programs were developed to reinforce the effect of light and sense of tactility of the surface, the latter was necessary to compensate for the "dryness" of extreme perfection of computer drawing and the hard paint finish needed for permanence. (#1 and #3 and background)

From a bird's eye view one zooms into **Human Scale**, Window #3. The sepia tones set it apart from the rest of the mural and evoke the patina of the past: sand dunes, archeological ruins and old maps. Here there is no single viewing point but a focus at the center of the mural on human activity, a sports group, moving out to greater spectacles in the stadium. Salt Lake City is seen from a satellite and is contrasted with ancient Rome, medieval Paris and 17th century London. Their forms have derived from being located on major rivers while Salt Lake City's grid pattern is framed in the ruins of Chan Chan a prehistoric Peruvian site, a desert city one sees through time.

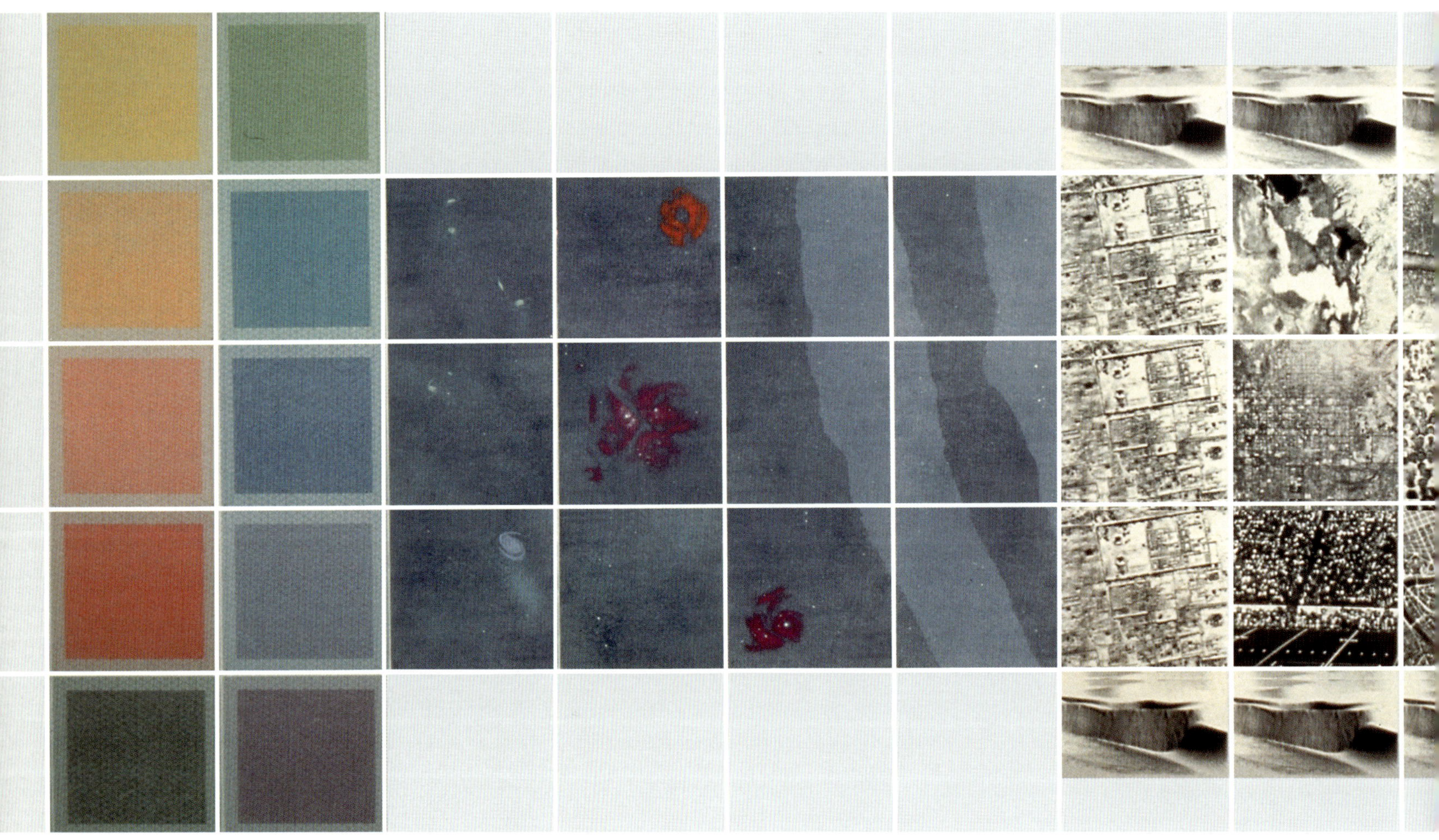

WINDOWS
1989 - 90, ALKYD ON ENAMELED STEEL, 8' X 30', COMPUTER BASED MURAL FOR THE
DATA PROCESSING CENTER, UTAH STATE CAPITOL. UTAH ARTS COUNCIL PUBLIC ARTS PROGRAM.

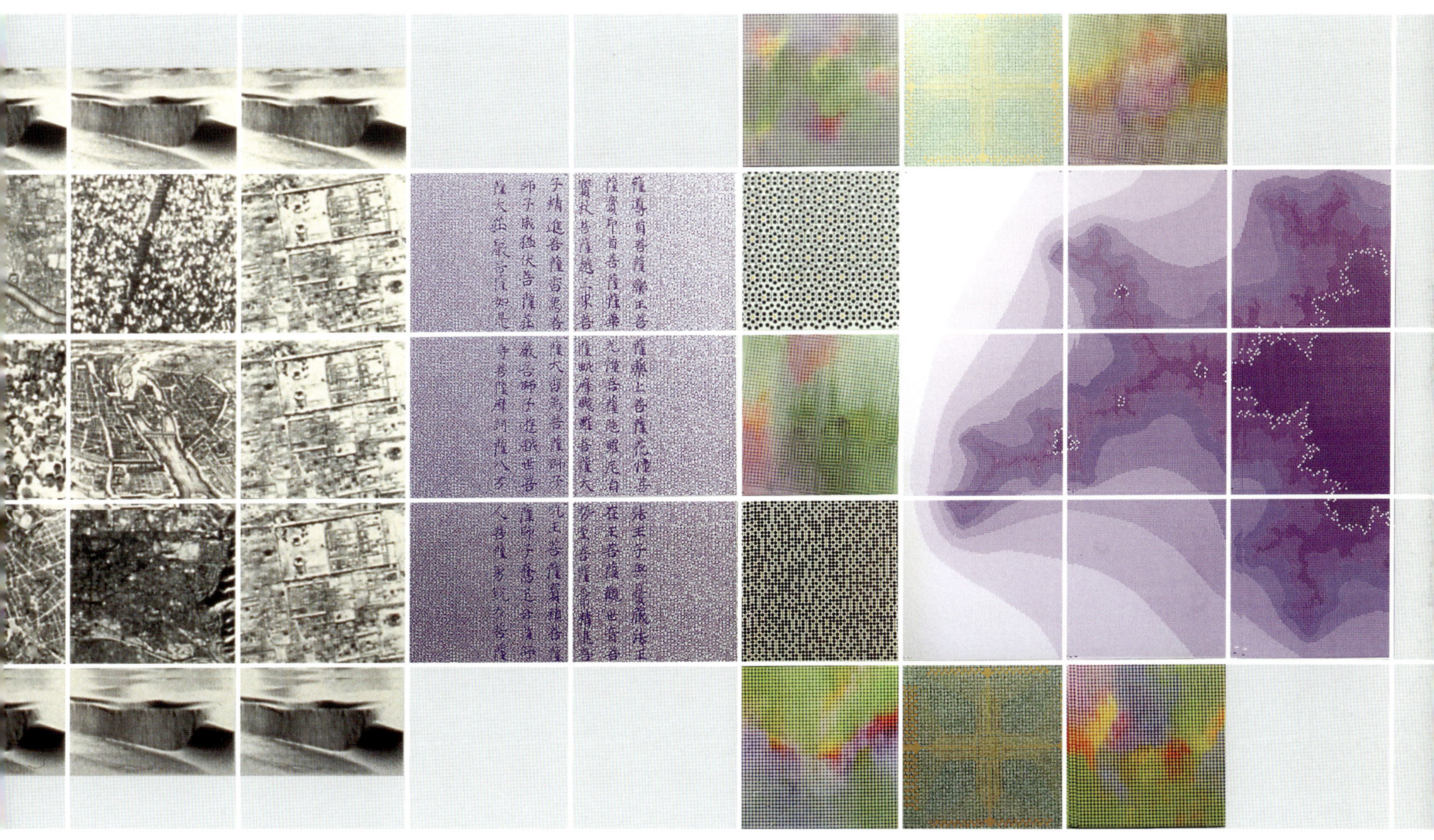

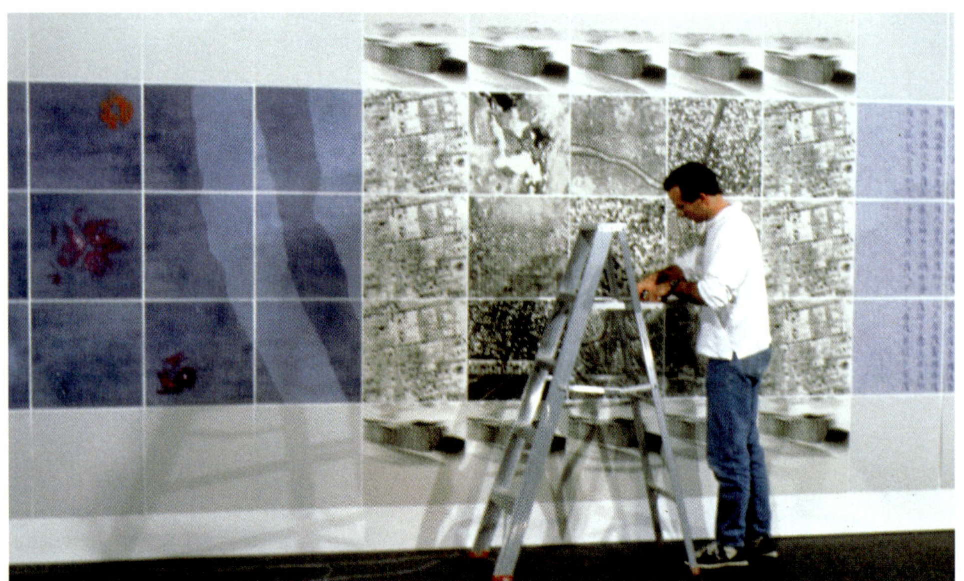

STUDIO ASSISTANT, ROBERT LIGHTY, INSTALLING WINDOWS IN THE ARTIST'S STUDIO

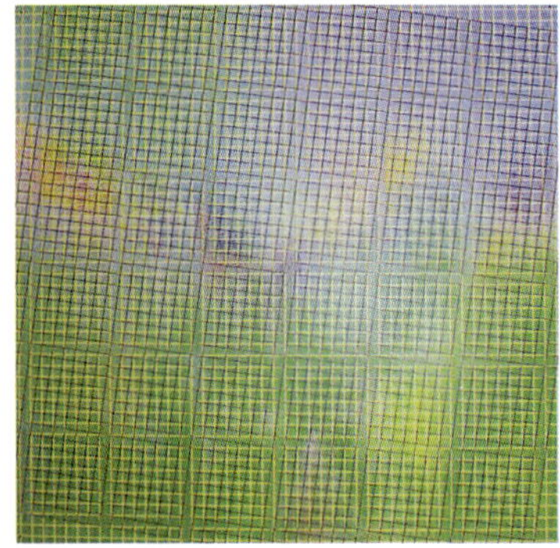

MEMORY STUDY FOR WINDOWS 1989, ALKYD ON ENAMELED STEEL, 17 13/16" X 18".
COLLECTION: TED & YEIKO NAGATA

Communication, Window #4, with its passage from the Lotus Sutra, provides a pause, visual and ethical between a ceremonial site in Peru and the Micro World of the 21st century. This famous Buddhist sutra urges politicians to practice ethical behavior, appropriate for its location near the meeting place of legislators. The hand calligraphy of ancient Chinese characters contrasts with the blandness of alphanumeric computer printing.

Micro World, window #5 is primarily a free interpretation of the memory storage of the computer: an early magnetic core structure recalls the abacus, an image of a computer chip in enlarged form suggests an Indian sand painting and a silicon molecule with pseudo coloring as seen through a scanning-tunneling microscope. For more current chips light refraction was suggested in the endless repetition of overlapping grids. "With the rapid changes in computer memory structure, I felt the interpretation should be suggestive rather than literal."

Fractalscape, Window #6 is rich in multiple meanings. It looks to the future and to the science of chaos. For the artist it was the most characteristic of the mathematical concepts the computer can implement and artistically the most labor intensive to execute. The movement of the mural from left to right needed a strong form to contain it. The Dendrite arms (a variation of the Mandelbrot set) provide a counter movement that leads the eye back from right to left. The mathematical concept was freely interpreted and by changes of the color contours conveys a feeling of land and sea. Here one is reminded of the OMIZU SERIES but in a hi-tech context.

It is one of those rare moments in the life of a work of art when the artist meets the poet within and the idea becomes experience. From lifelong exploration of color to the visual poetry of a screened Fractalscape in outer space, Anna Campbell Bliss is a navigator in touch with the boundless possibilities of "the great tradition of artistic experimentation."

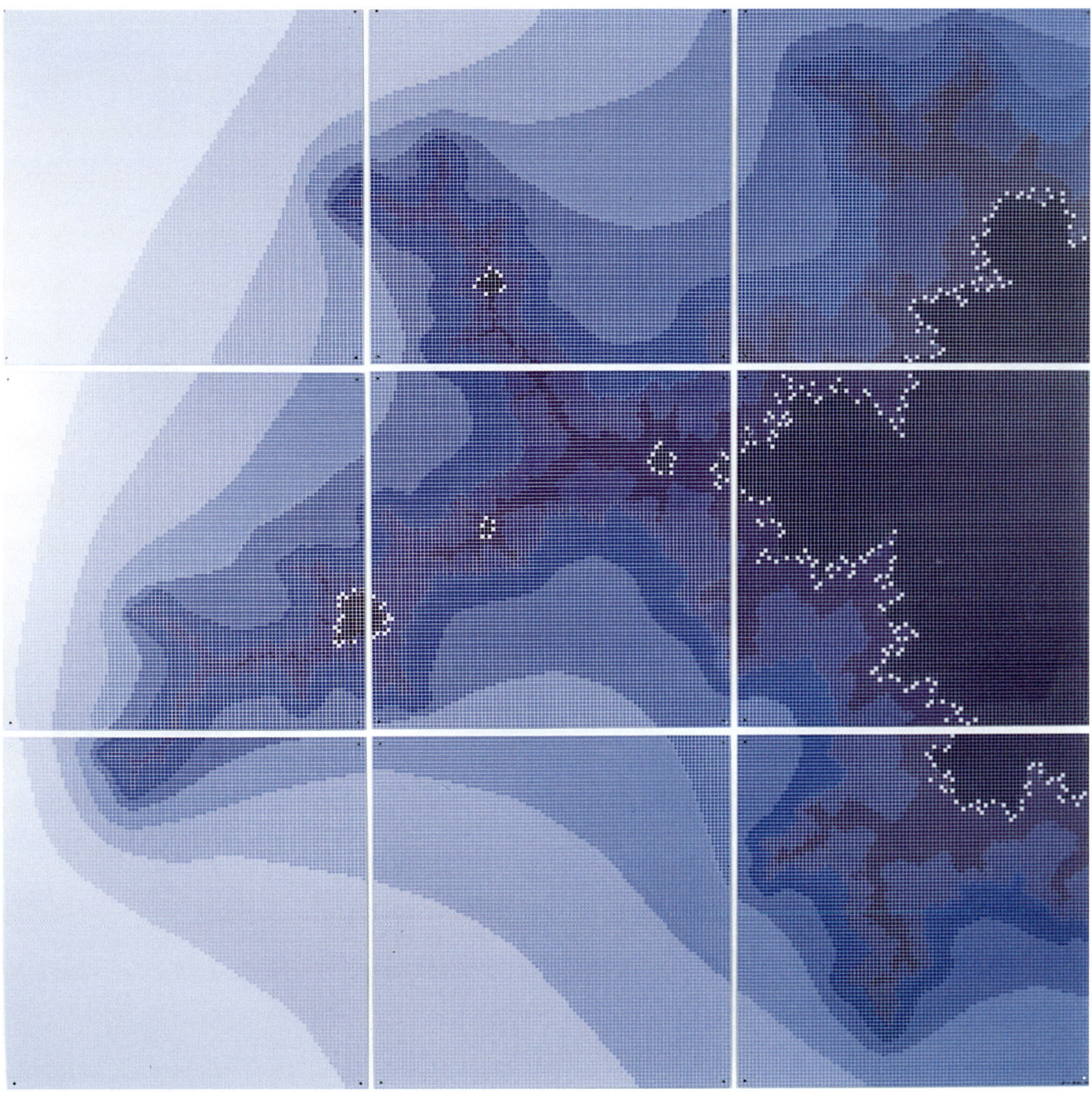

FRACTALSCAPE

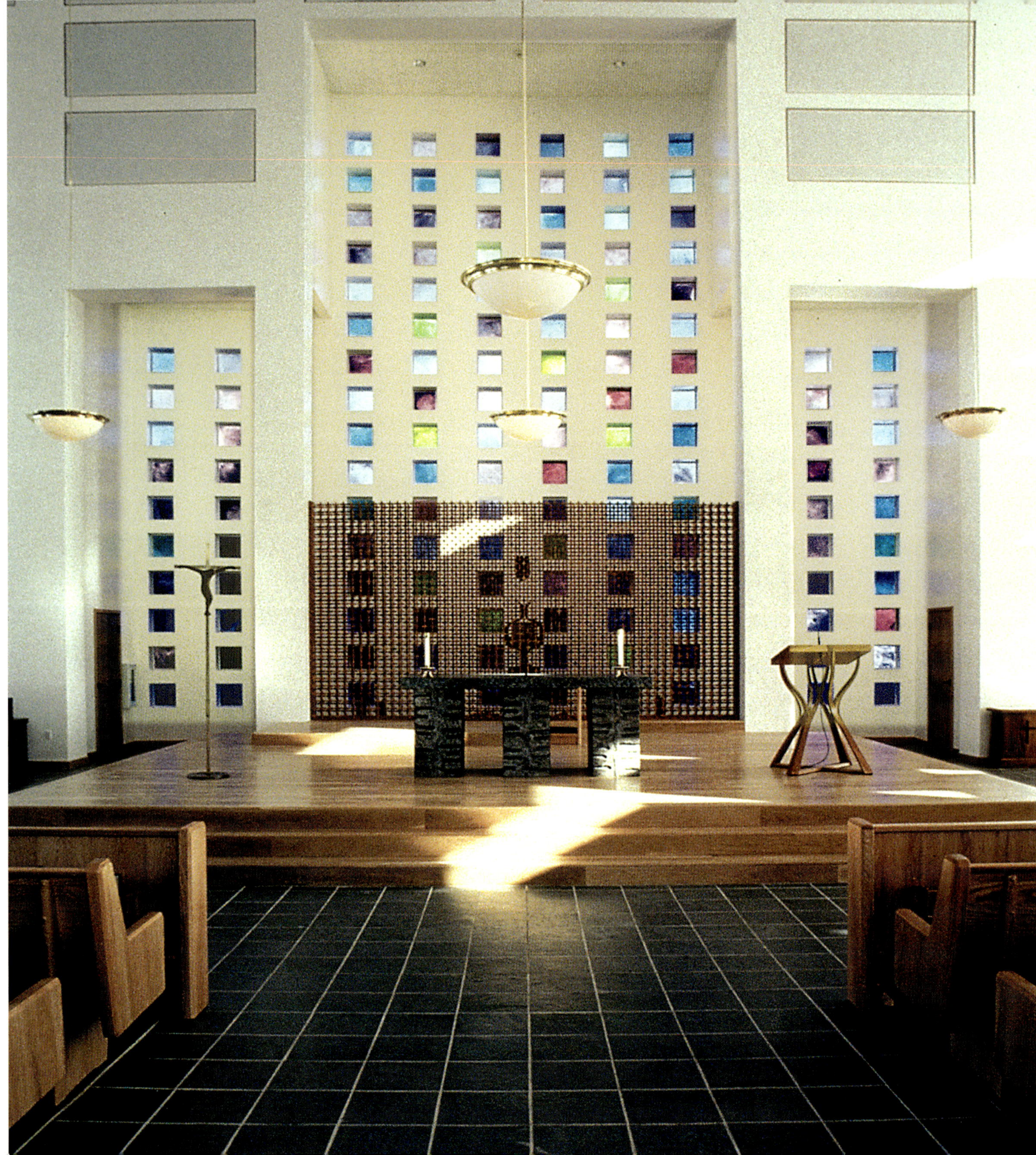

LIGHT OF GRACE

Following WINDOWS Bliss was asked to consult on the art and design for the church, bringing together artists for the project and to create the glass window wall. The natural light during the day needed to be modulated behind the altar. The architects, Michael Stransky and David Brems, planned a simple, low cost building using natural light and materials and well sited for views of the mountain environment. "As we approached the art and interior we had a strong concept to work with and it became important to develop its uniqueness while respecting Catholic tradition."

The cast glass by Heritage Glass of Logan, Utah was made by a process that has changed little since the Middle Ages. The molten glass is taken from the furnace in large ladles, poured into molds 1'' thick and then gradually cooled. Imperfections in color mixing, bubbles, etc. are part of the process and add to the beauty of the richly saturated color. The color is roughly graded transmitting 80% of the light in the upper rows and decreasing to 40% in the lower areas giving more intensely saturated color. The design is simply fingers of light focusing on the altar.

Inside the church's white interior, light creates an array of color vibrations and reflections. Outside on snow, the dance goes on. "The church building performs many functions but important to me was creating a sense of reverence and respect without detracting from the joyous quality of the light. Although the budget was minimal, the effect of the work is very powerful. The color changes with the sun and seasons provides endless interest and a sense of celebration while recalling the church's great tradition of stained glass." The ambiance created offers a continuing source of beauty and inspiration.

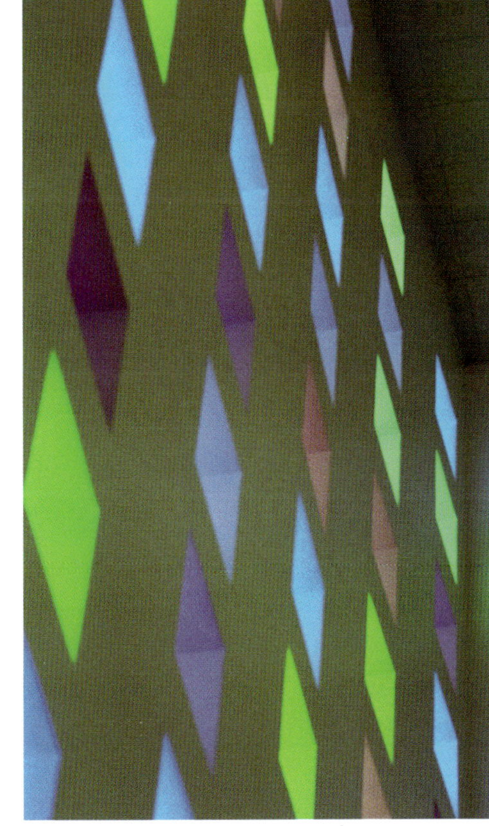

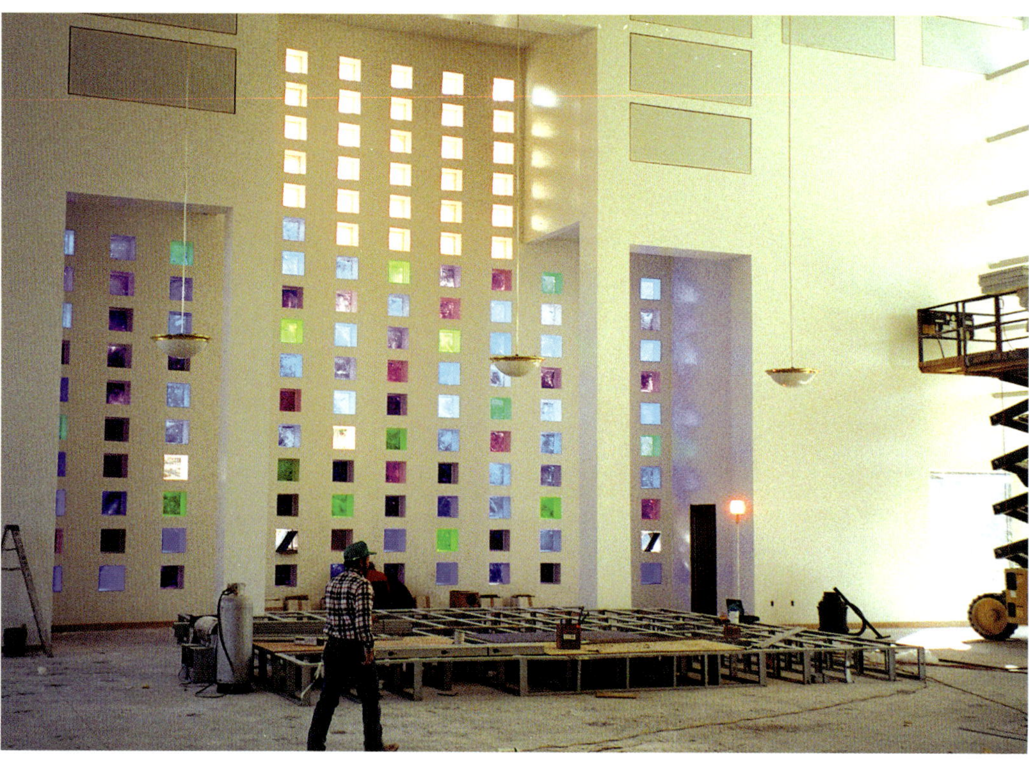

Installation photo, **Light of Grace**

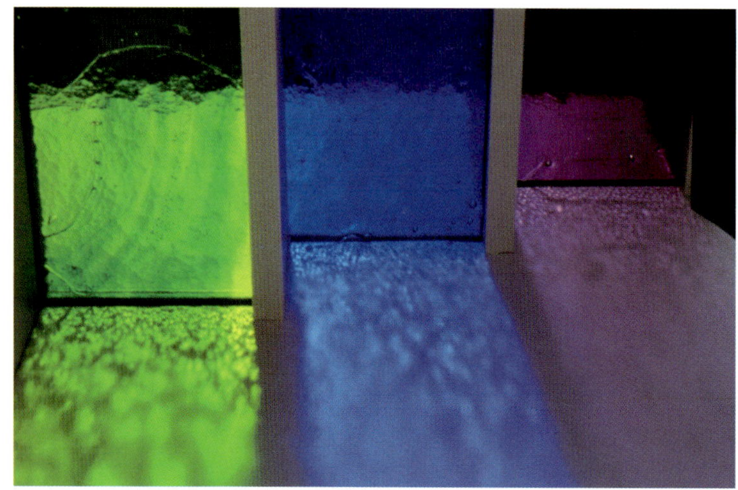

Studio study for cast glass

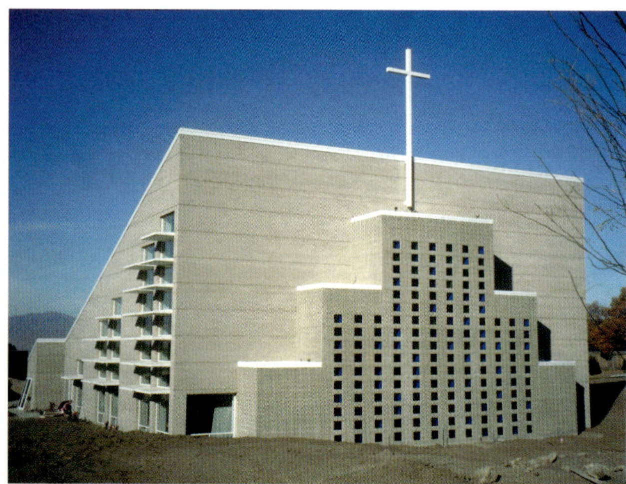

Exterior view, St. Thomas More Catholic Church

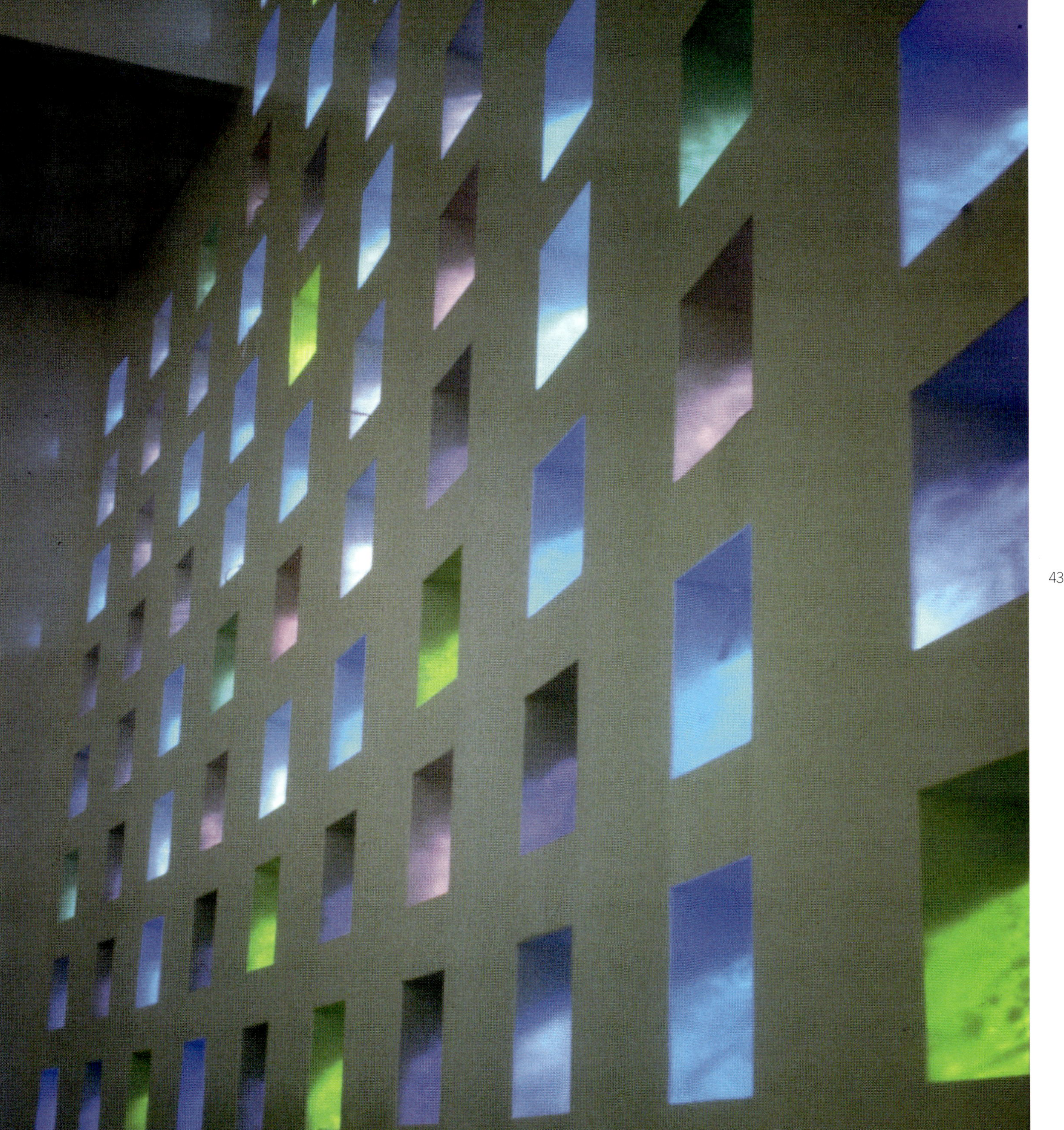

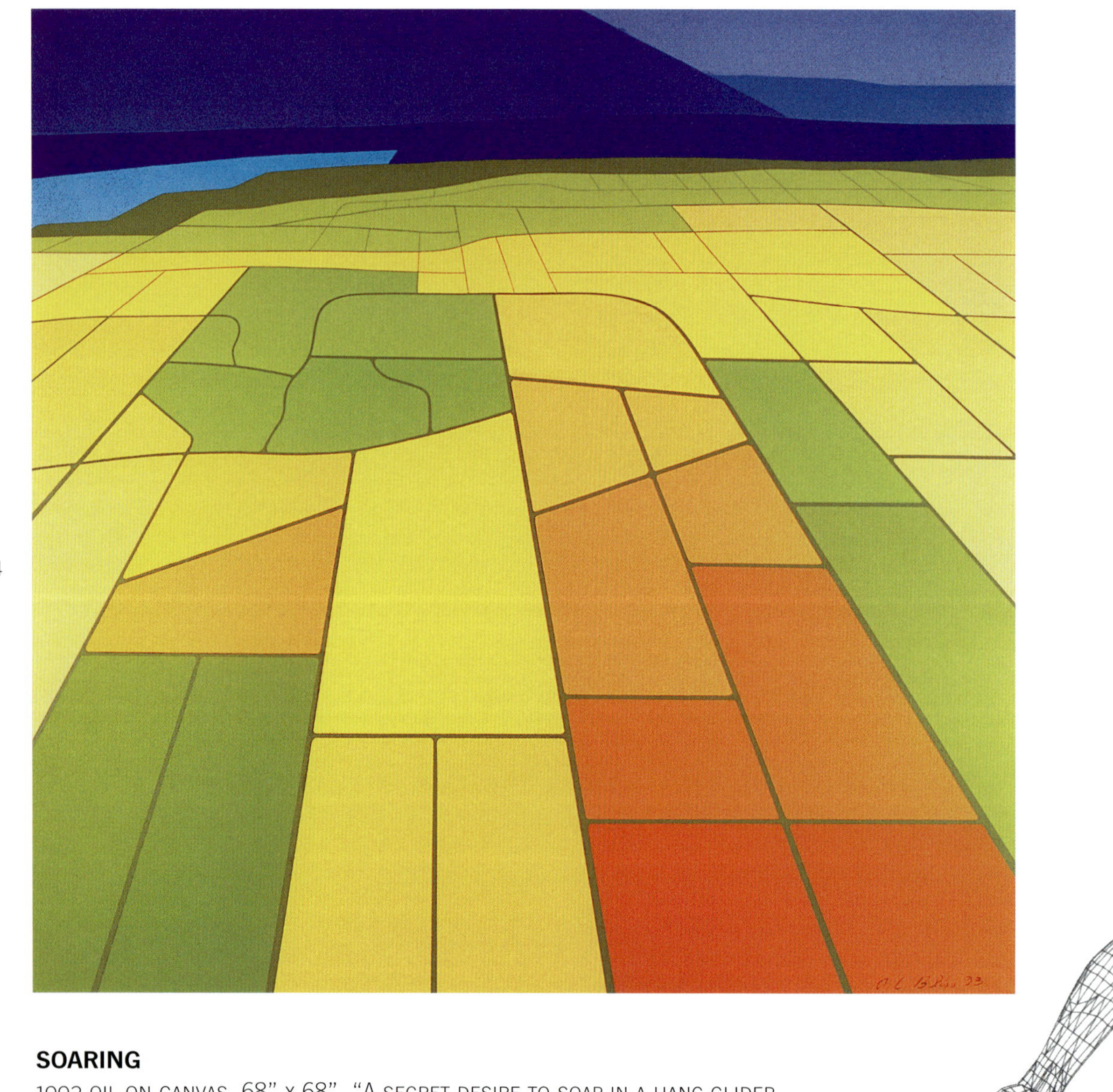

SOARING

1993 OIL ON CANVAS, 68" x 68". "A SECRET DESIRE TO SOAR IN A HANG GLIDER INSPIRED THE WORK AND FASCINATION WITH THE PATTERNS OF CULTIVATED FIELDS SEEN FROM THE AIR. MANY HAVE CLAIMED THE PAINTING AS A FAMILIAR VIEW OF THEIR VALLEY, FROM ILLINOIS TO SACRAMENTO, BUT IT IS NON LITERAL IN CONCEPT."
COLLECTION: WILLIAM & BETSY CAMPBELL

DISCOVERERS

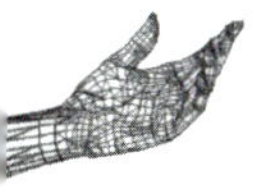

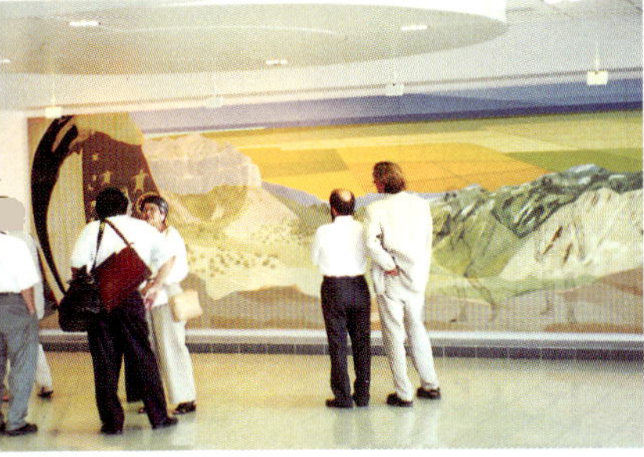

SALT LAKE CITY INTERNATIONAL AIRPORT

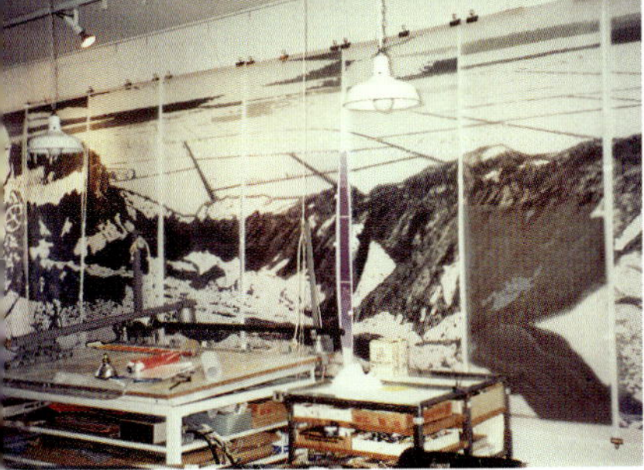

STUDIO INSTALLATION STUDY

Coming to Utah from a very different environment, Bliss was attracted by the great beauty and variety of the landscape and photographed it extensively. For the Airport commission she celebrated "the extremes one finds in this area from the natural beauty of the wilderness to highly cultivated lands and from the Indian art of Utah's prehistoric past to the pioneering computer and hi-tech developments that are transforming life here."

She describes the mural "as a fantasy with more than one level of reality. The central area is the austere wilderness the DISCOVERERS found, the upper portion their dreams of transforming it. The outer prehistoric area and rock formations are overlaid by a computer screen and wire mesh figures welding past and future."

What her words do not convey is the sensation of the three figures emerging as one approaches the mural from the "D" concourse. The figures, universal in concept, are the artist's computer creation. Standing before them images reverberate in my mind stirring up an art historical past. The figures gesture to markings engraved on a tomb, "ET IN ARCADIA EGO", a 17th century neo-classical painting by Nicolas Poussin. Ages echo through this phrase reaching back to Virgil's elegaic meditations and forward to EXTENDED VISION.

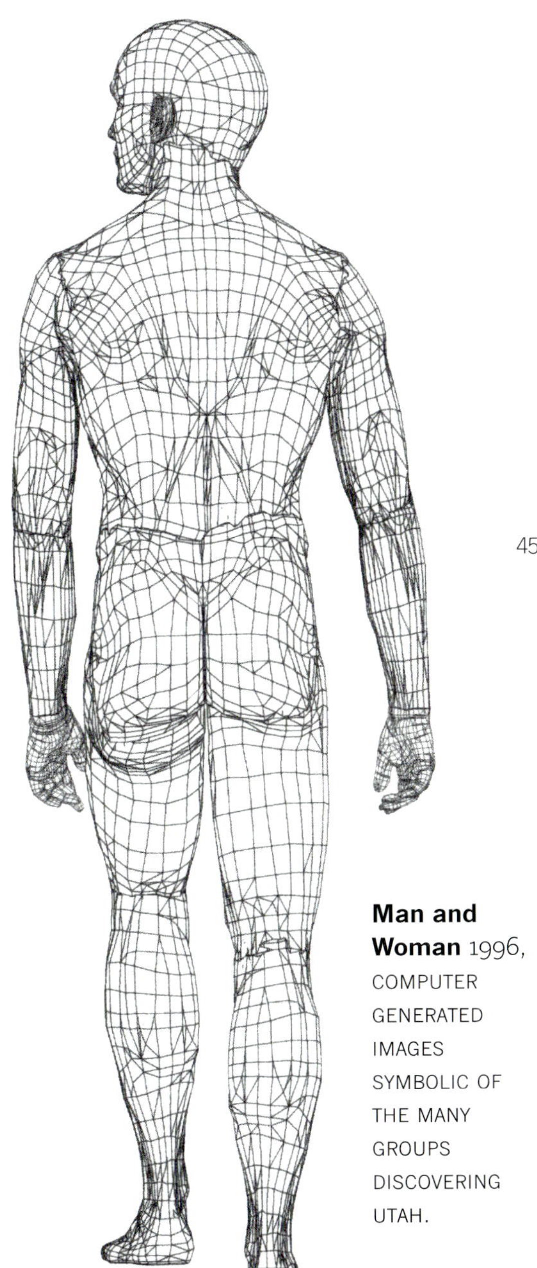

Man and Woman 1996, COMPUTER GENERATED IMAGES SYMBOLIC OF THE MANY GROUPS DISCOVERING UTAH.

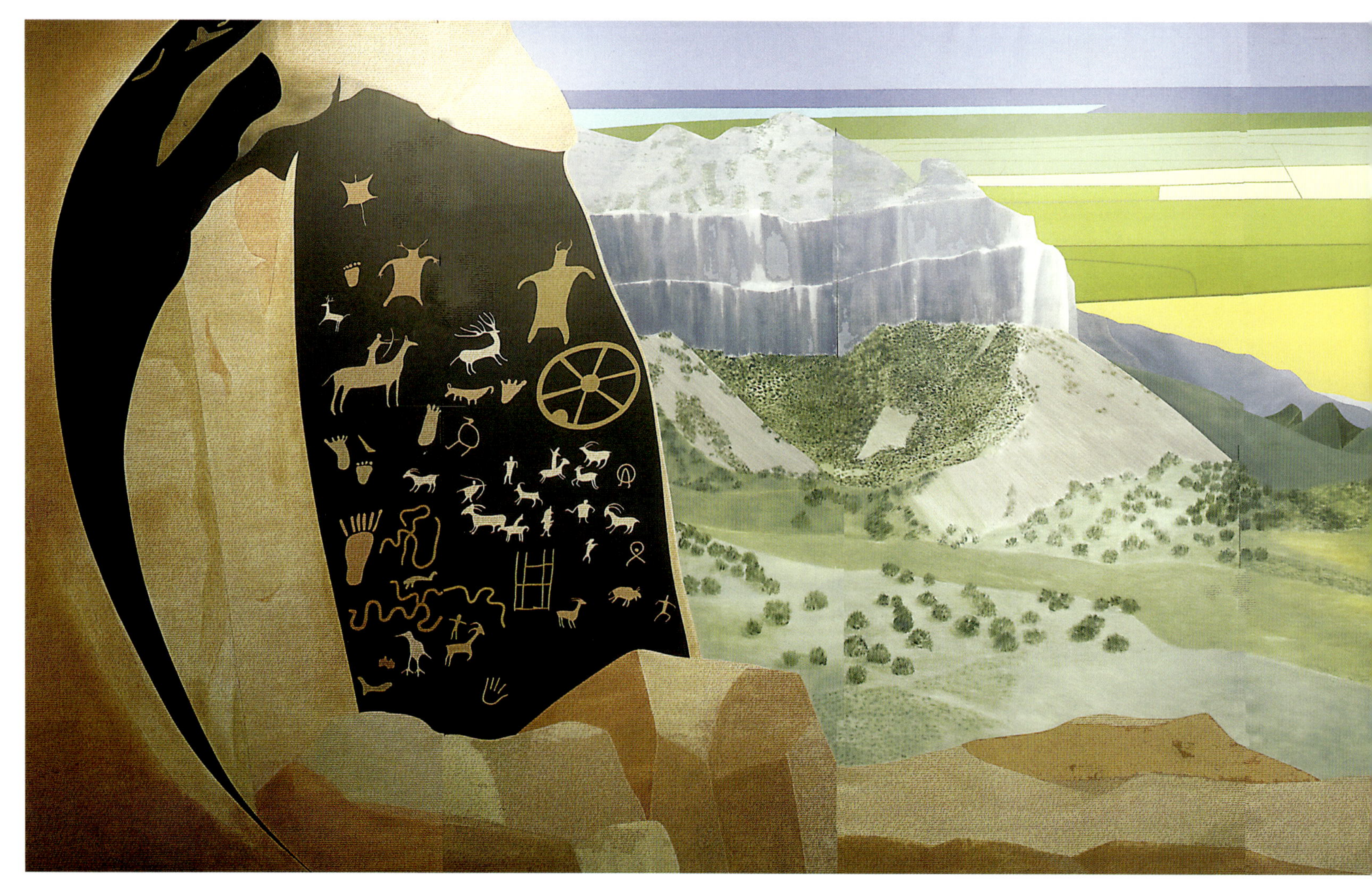

DISCOVERERS

1996 OIL ON CANVAS 8' X 25'-6" WITH SCREENED
COMPUTER BASED IMAGES,
SALT LAKE CITY AIRPORT COMPETITION

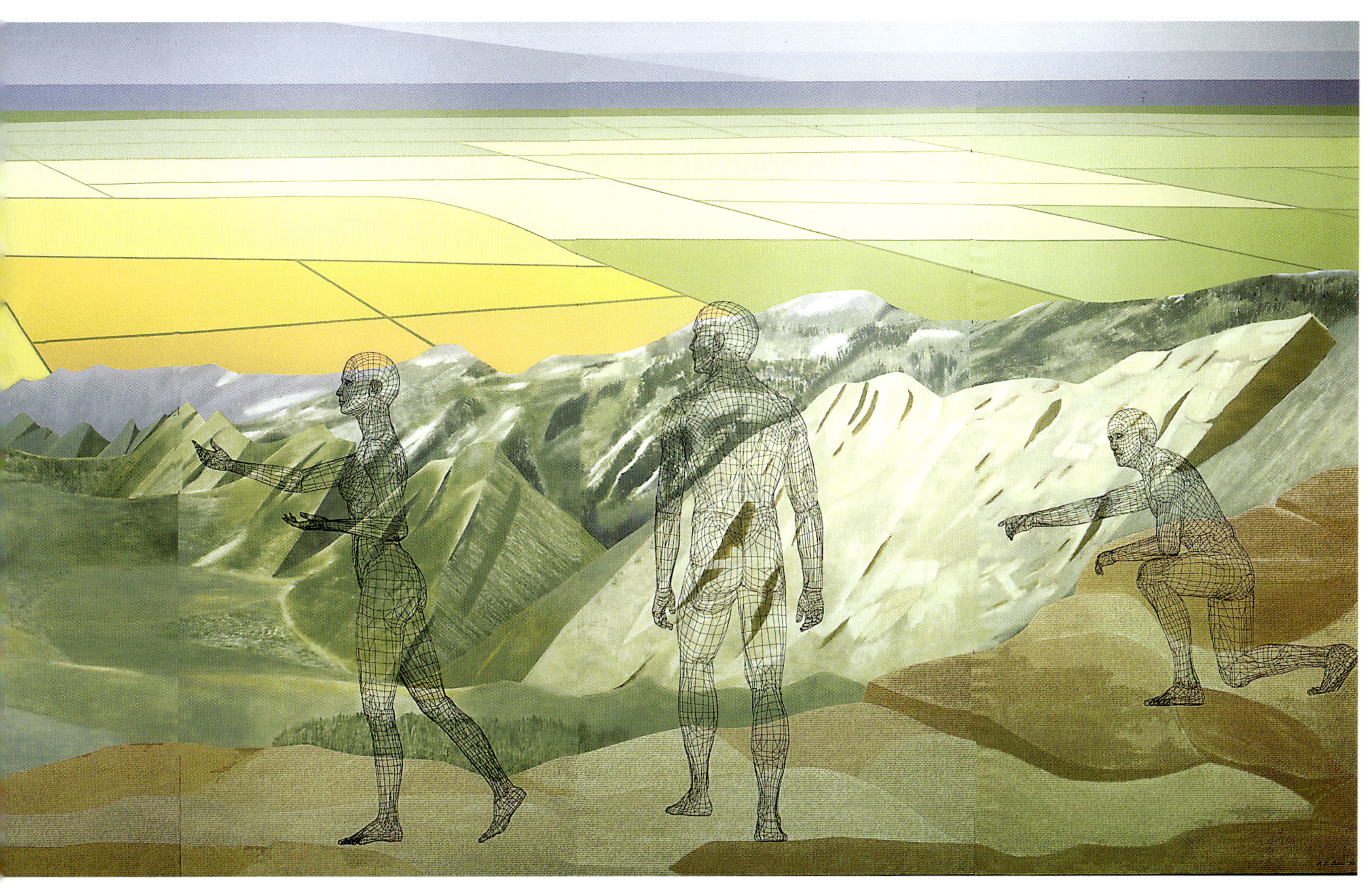

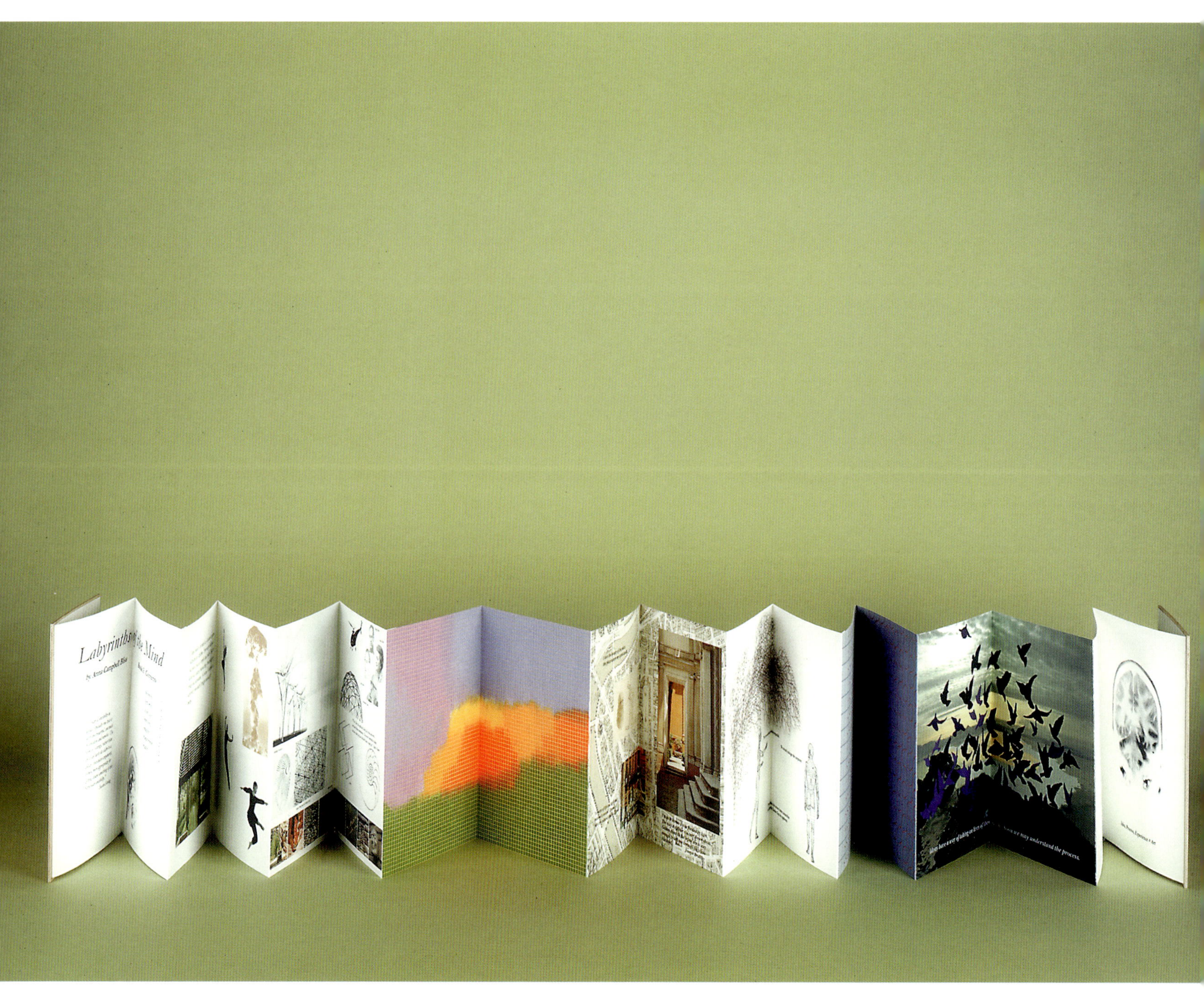

LABYRINTHS OF THE MIND

1998 – 2000
REVISED FOR EDITION OF 50,
A JAPANESE FOLD BOOK,
MIXED MEDIA: DIGITAL PRINTING,
PAINTING, SCREEN PRINTING,
WITH CENTER PAGE UNIQUE
FOR EACH BOOK.

Knowing her interest in the Book Arts, Madelyn Garrett challenged Anna Bliss to do a book for the competition and exhibition, Westward Bound. She was surprised and pleased to have her first book accepted. Concepts of mind and recent brain research inspired exploration of her own experience. Particularly interesting to her were the visual, verbal and tactile references that trigger memory.

The introduction by Susan Sontag is the best explanation of what she was trying to do and the type of book that was created. The page titles provide clues to the content which is more abstract and visual than a typical biographical account. Her interest was conveying ideas by visual means with the minimum of text.

With elegant economy, the artist invites us into the garden of her mind. This is her map. Slipped from its brilliantly colored silk jacket. A cover of pale Baltic birch holds a "fold book" that opens in seven sections like a Japanese screen.

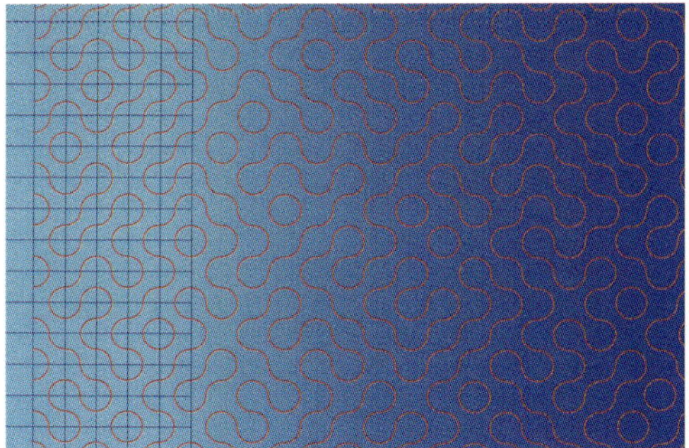

Becoming reflects the emergence from the security of childhood into the adult world.

Exploring introduces the many sources of intellectual and visual inspiration that have influenced her life. The color photos record her interest in archeology and the physical explorations of Peru, Bolivia and southern Utah.

Color/Light: Each page is an original hand painted and screened work. It was included as the focus of her art and research and also for contrast with digital printing which has less color saturation and lacks tactile qualities.

Enriching records her fellowship in Rome which came at a very critical time in her life. Anna looks upon that city as her spiritual home.

Imaging refers to her long experimentation and frequent frustration with the computer as an artistic tool.

Transforming was added to the revised version to reflect some of her programming interests. It is based on the Truchet tile algorithm.

Creating introduces her photography in a constructed image to convey the way that ideas take off and create their own demands.

LEFT, TRANSFORMING

I often think of the mind as a luxuriant garden, cultivated for a lifetime. From early childhood it has been a place of comfort and shelter. But the world has a way of encroaching, threatening my refuge and inspiration. In dreams it becomes a labyrinth from which I struggle anxiously to emerge.

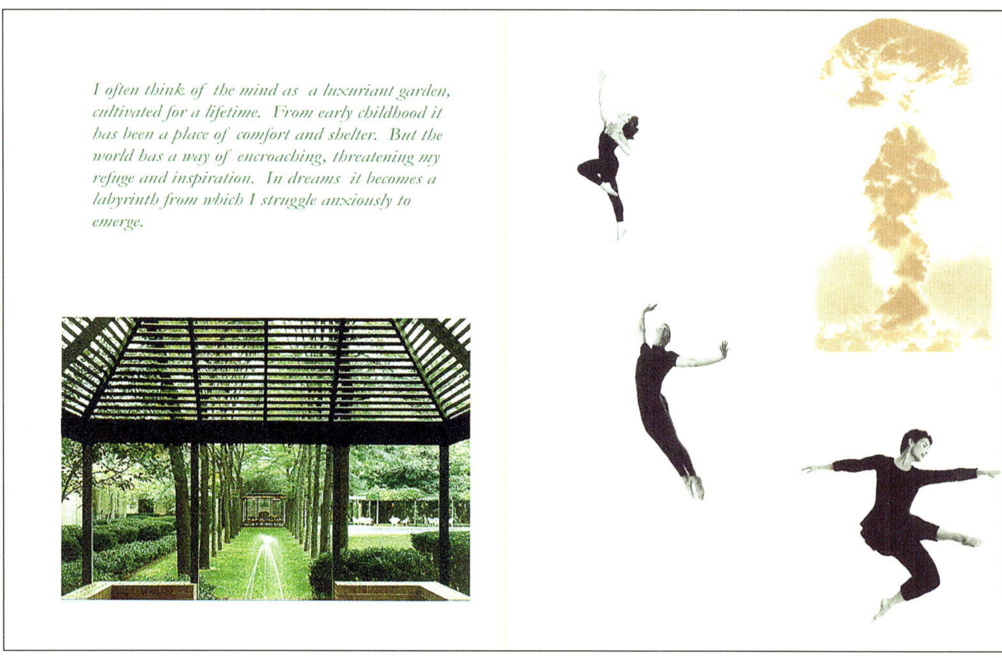

BECOMING

EXPLORING

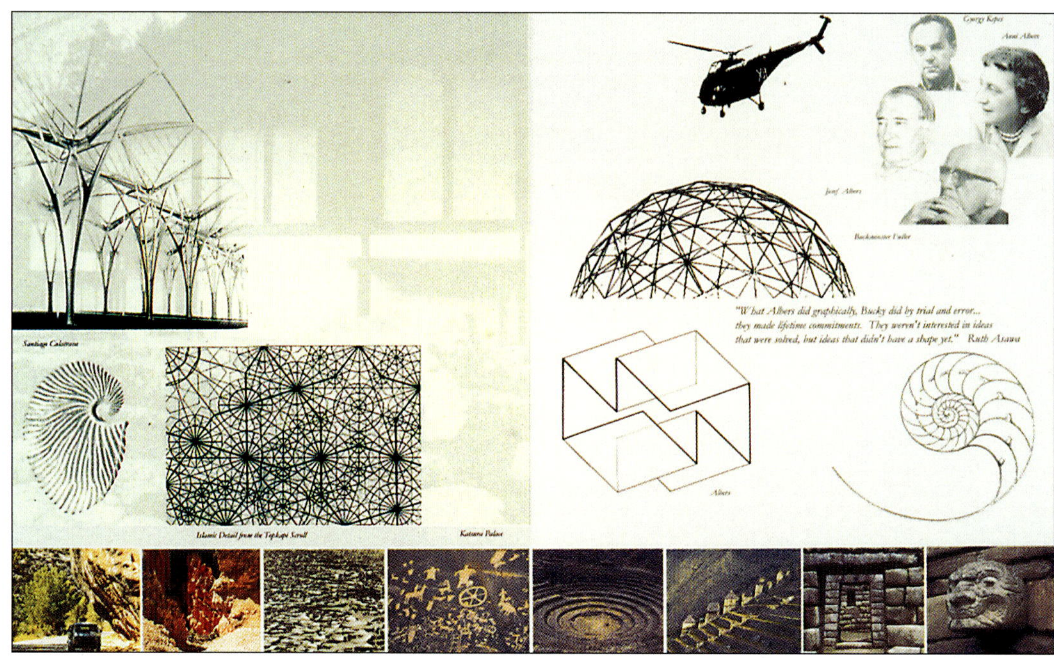

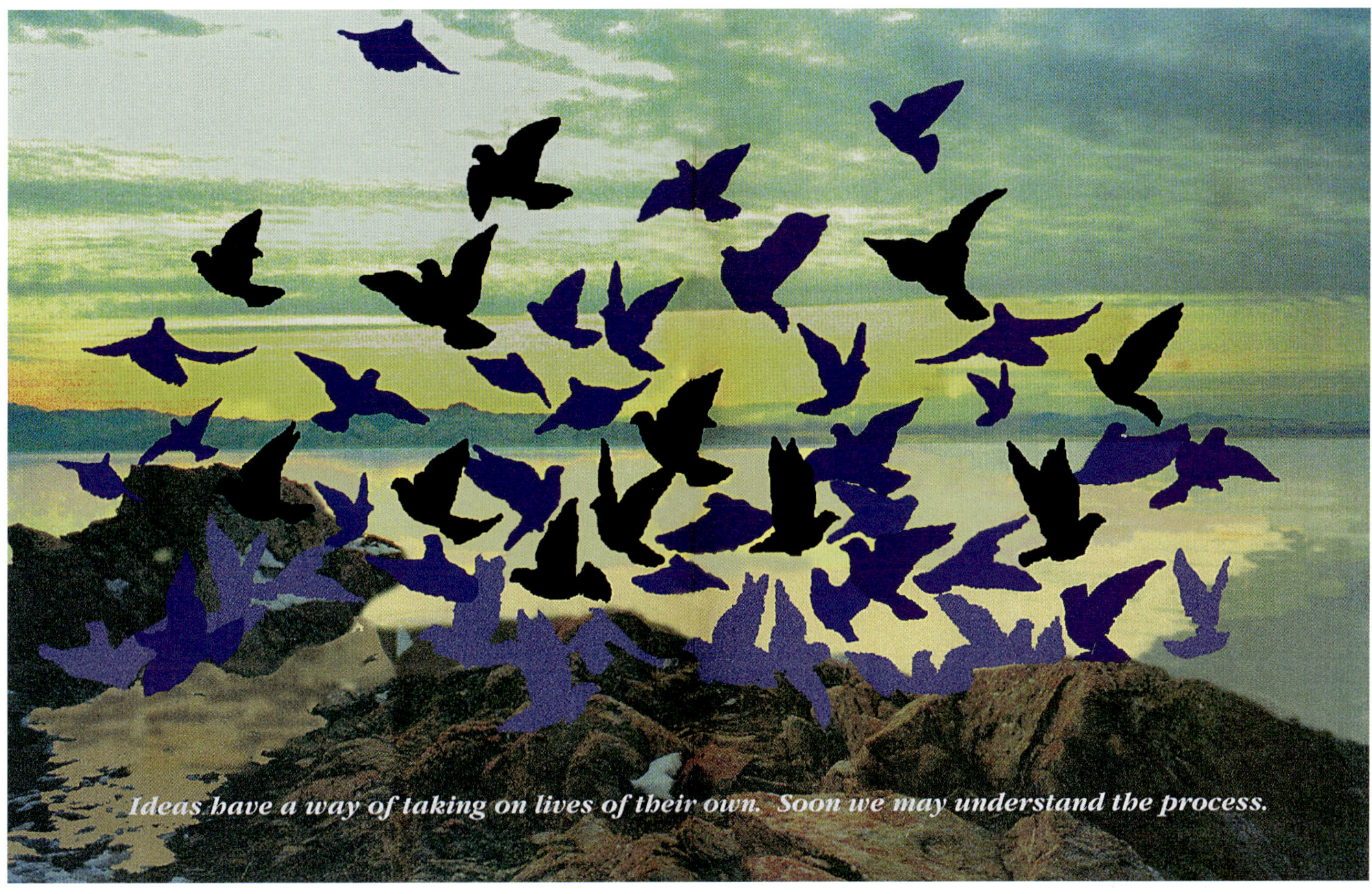

Like chapters in her book, Anna's art is suggestive rather than explicit: "jottings of a sensibility" in a state of becoming. So the book unfolds, the water ripples Lake Truchet and the birds fly above Antelope Island. In the center of the "garden" an original colorscape emerges from a screened grid, "the touch that disturbs both shin and gyo," that transforms her life into art:

"to snare a sensibility
alive and powerful . . .
tentative and nimble"

excerpts from
Susan Sontag

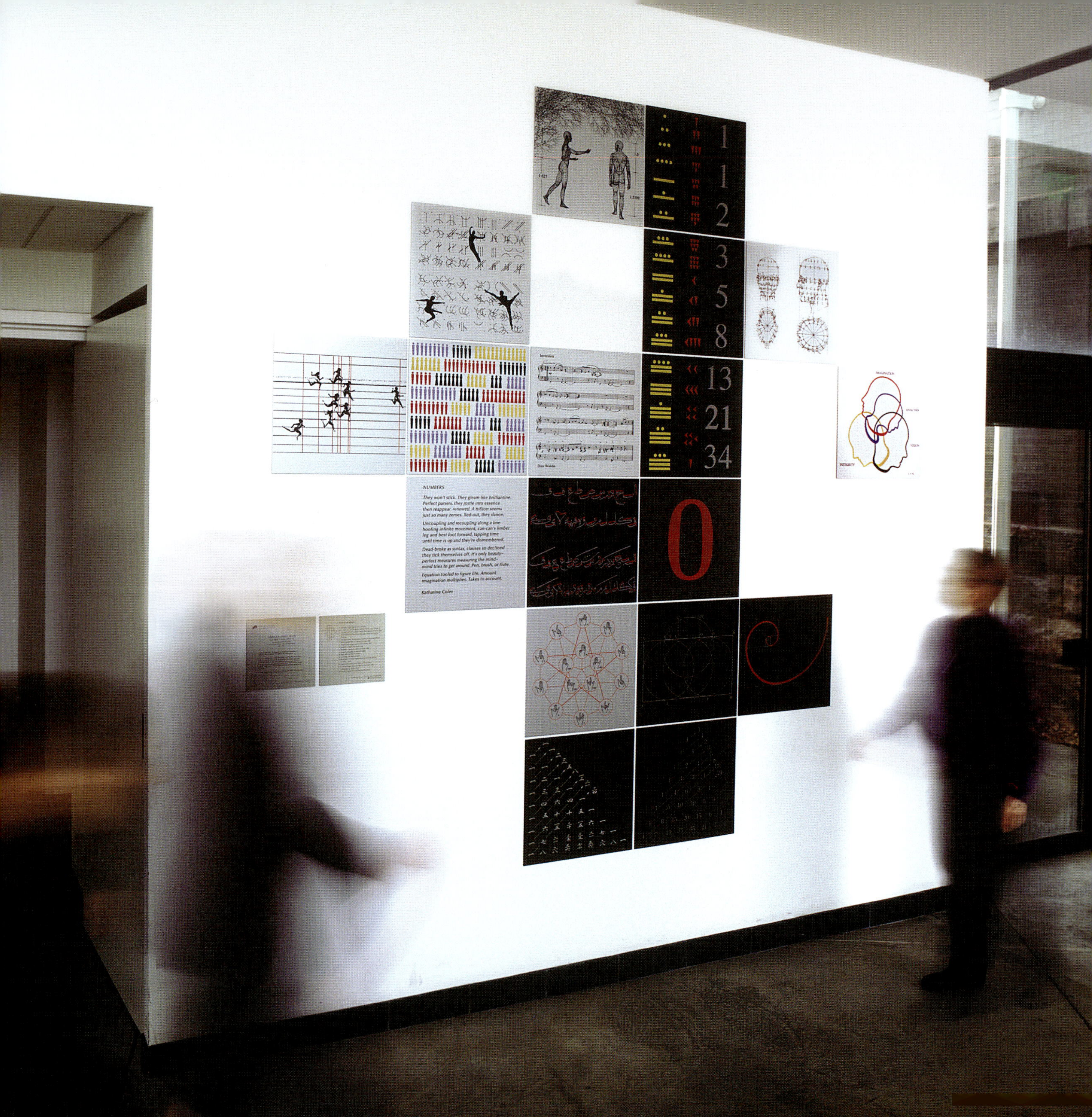

NUMBERS

They won't stick. They gleam like brilliantine.
Perfect parsers, they jostle into essence,
then reappear, renewed. A trillion seems
just so many zeroes. Xed-out, they dance,

Uncoupling and recoupling along a line
booting infinite movement, can-can's limber
leg and best foot forward, tapping time
until time is up and they're dismembered.

Dead-broke as syntax, clauses so declined
they tick themselves off. It's only beauty–
perfect measures measuring the mind–
mind tries to get around. Pen, brush, or flute.

Equation tooled to figure life. Amount
imagination multiplies. Takes to account.

Katharine Coles

1
1
2
3
5
8
13
21
34

0

Invention

IMAGINATION
EXALTED
VISION
INTEGRITY

EXTENDED VISION

2001-03, MIXED MEDIA, LASER ETCHED AND SCREEN PRINTED ON 100 18" X 18" ANODYZED ALUMINUM PANELS EXTENDING ON THREE FLOORS OF THE COWLES MATHEMATICS BUILDING, UNIVERSITY OF UTAH. AN ART COMMISSION OF THE UTAH ARTS COUNCIL PUBLIC ARTS PROGRAM.

In a national competition, the faculty asked for art for their building expansion that would create the sense that this is a "house of mathematics" and convey its wide ranging influence. They also asked that it be a source of pleasure, intrigue and inspiration for future generations, a rare statement from a building committee.

This was a challenge that demanded new solutions. A solo work on one floor did not satisfy the stated goals. "I visualized a multifaceted work that extended throughout the building to develop structural connections with the sciences, arts and culture while suggesting the range and beauty of mathematics. It could be experienced as individual works of art, in related clusters and as a total work of art. Organizing the plates in clusters suggested the openness of ideas and the possibility of extending them.

The lobbies on each of three floors became logical focal points for major themes:

1 – **Numbers and Measure** – with corridor extensions including Islamic and Greek concepts and later geometric studies.

2 - **Intersections** – mathematics and nature, major directions in the sciences and a third group from ethnic designs and early structures to major architectural works.

3 - **Outer Space** – and terrestrial navigation before the instrumentation for longitude.

EXTENDED VISION , the largest of Bliss's site specific works, has absorbed her time and imagination for the last three years. In her focus on the major themes that ascend through the three lobbies of the new addition to the Cowles Mathematics Building, she travels through changing times, technologies and cultures. With the eye and mind of a Renaissance woman working with the tools of a computer age, she examines the intersections between art and science that have absorbed her for decades. On the 100 plates that extend vertically and horizontally through the building, the mosaic of her work weaves a global web that reveals the brilliance of Euclid, Darwin, Ptolemy, Einstein, DNA, RNA,

Through a Mayan count of yellow dots and bars Through red wedges of Babylonian cunciform Through the graceful sequence of the Fibonacci series.

Down the numbers go to the big red zero Within a random bitstream.

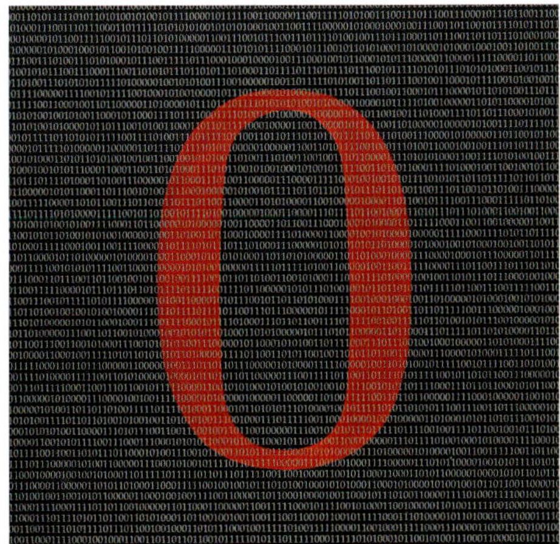

Modulars 2001
Classic ratio -
1:1.618

Numbers:
yellow- Mayan,
red- Babylonian,
grey- Fibonacci

Choreography by
Lucinda Childs,
RDT dancers &
Alvin Ailey

Numbers:
yellow- Mayan,
red- Babylonian,
grey- Fibonacci

Head studies by
Piero Della
Francesca, from
De Prospectiva
Pingendi

Olympic racers:
Swatch photo
courtesy of
Associated Press

Who Counts?
Who is Counting?
Census 2000

Original jazz
composition by
Dan Waldis, 2001

Numbers:
yellow- Mayan,
red- Babylonian,
grey- Fibonacci

Aspects of mind.
One equals four.

Numbers,
a poem by
Katherine Coles
2001

Islamic
calligraphy,
measured
studies

Zero,
with random
hit stream

Counting
with hand
gestures

Rennaissance
pictoral
organization
based on the
golden number -
1.618

Logarithmic spiral
using Fibonacci
proportions

Chinese triangle
showing
Fibonacci
numbers with
calligraphy by
Mishio Zushi

Pascal's
triangle
showing
Fibonacci
numbers

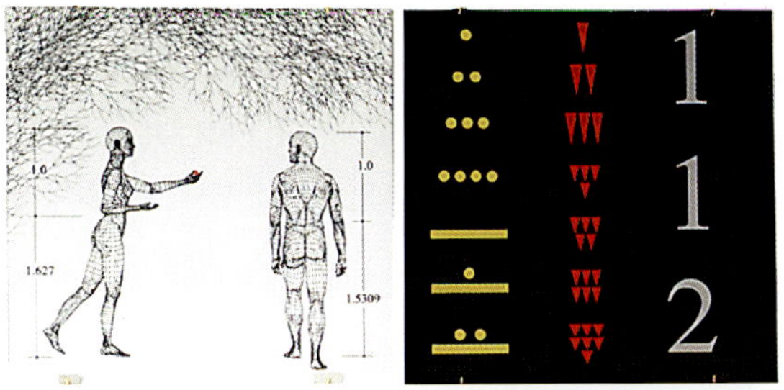

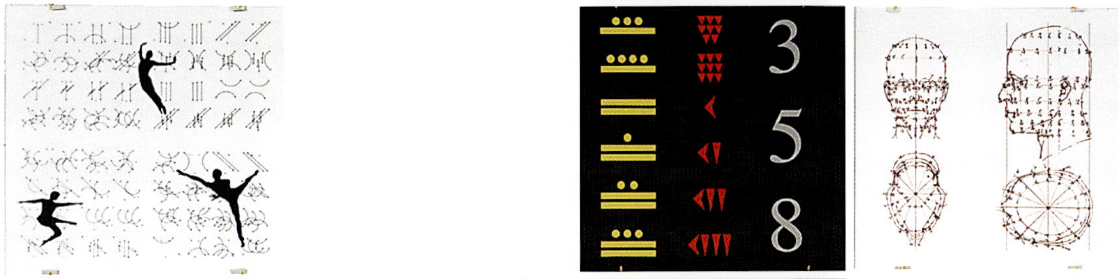

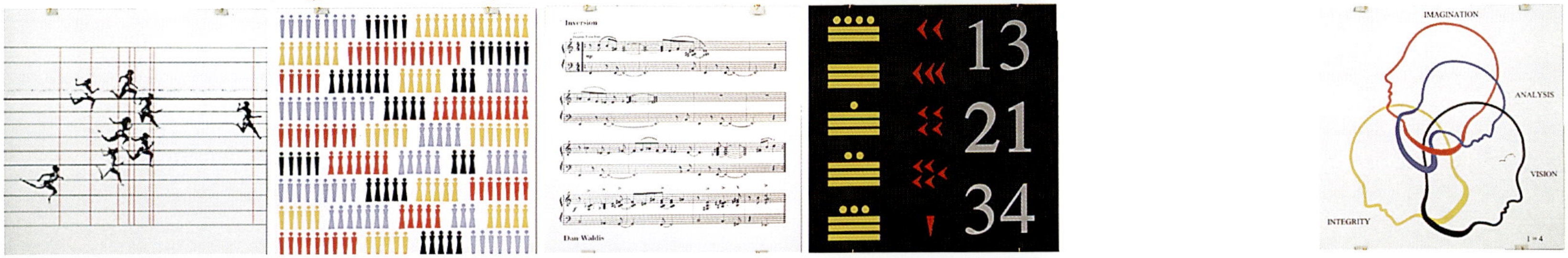

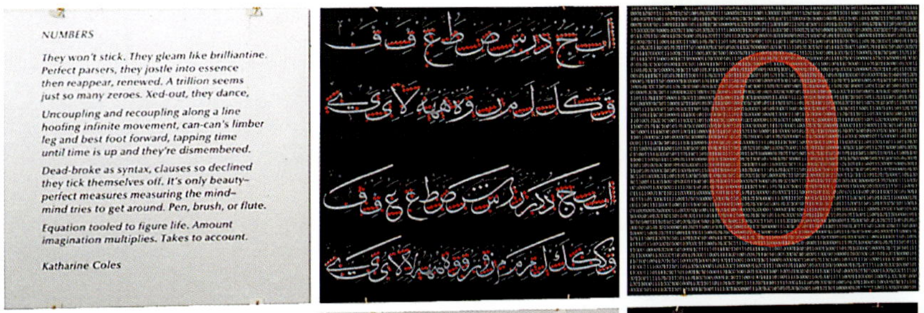

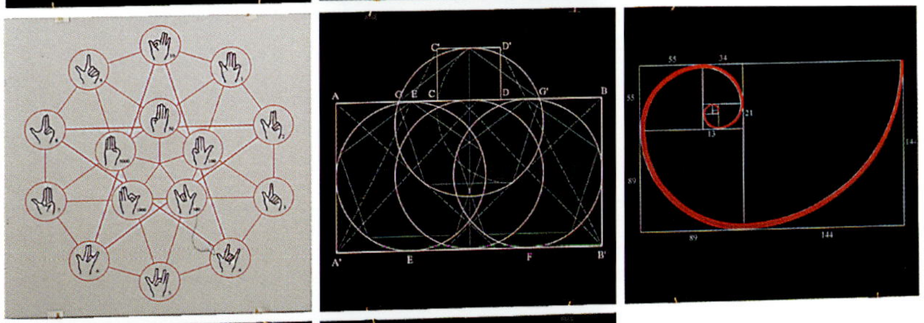

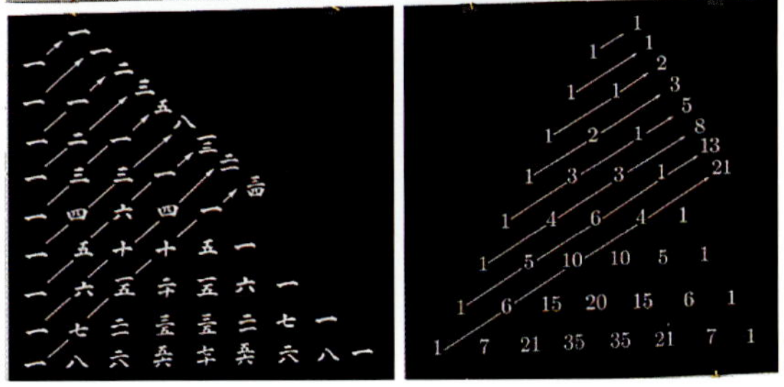

NUMBERS & MEASURE 2001

First Floor Lobby
18"x18" anodized aluminum plates
lazer etched black and screen printed

Cowles Mathematics Building
University of Utah
Salt Lake City, Utah

Utah Arts Council Public Arts Program

NUMBERS/MEASURE/GEOMETRY First Floor Corridor

13TH C. YAQUT AL-MUSTASIMI DESCRIBES
CALLIGRAPHY AS A SPIRITUAL GEOMETRY

ISLAMIC GRILL AND
TYPICAL PATTERN

EUCLID'S ELEMENTA GEOMETRIAE 1482

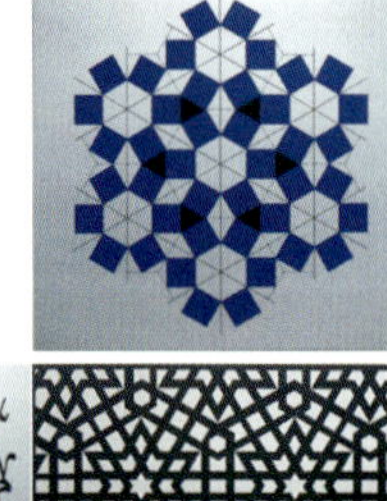

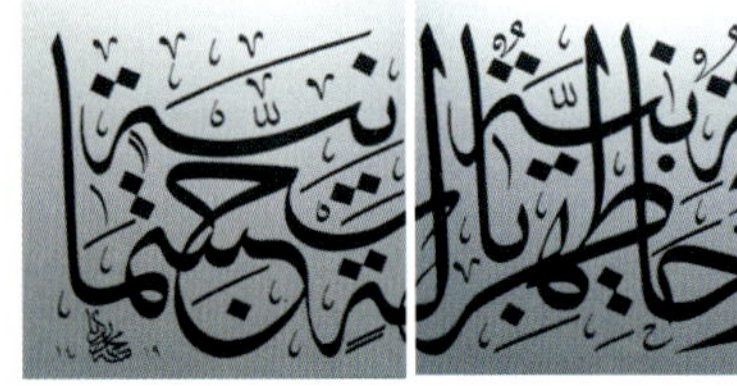

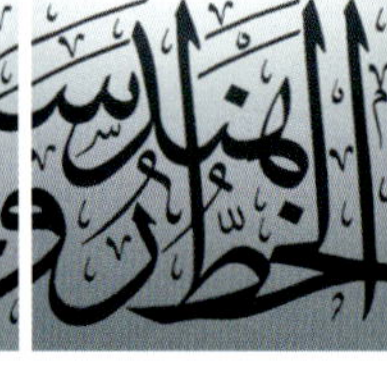

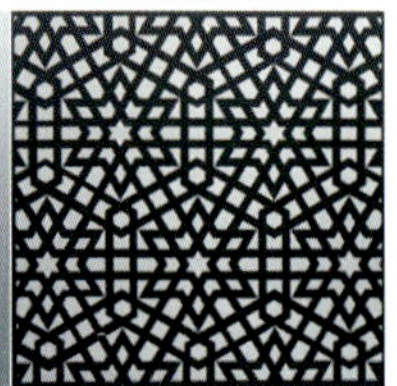

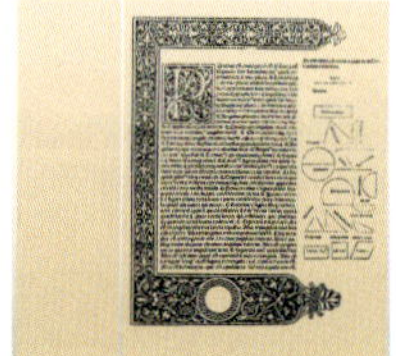

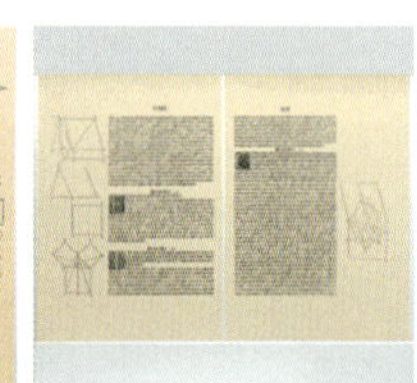

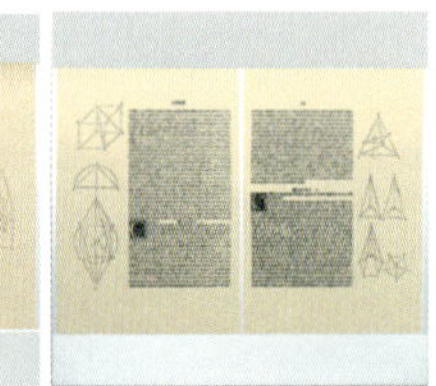

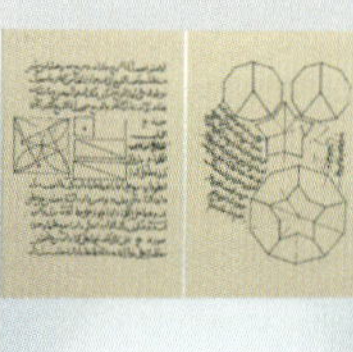

Moirè study   Earth's path with Einstein's equation   Hexagonal study   Spiral geometry Geodesic dome   Implied geometry   Harmonic growth   FASS curve with figure ground reversal

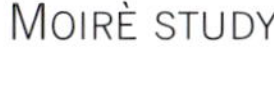

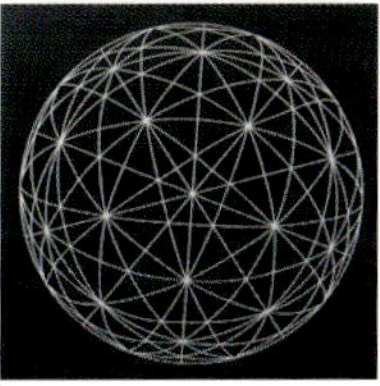
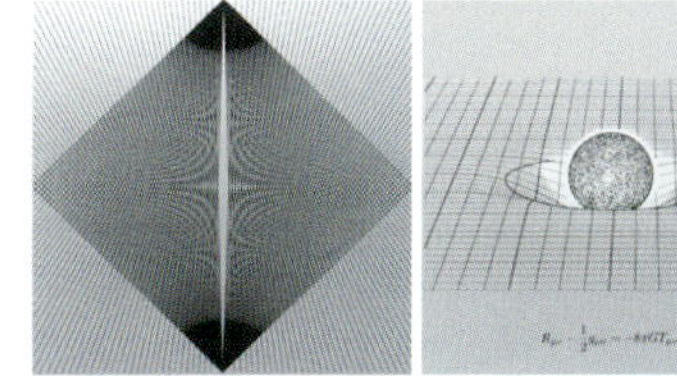
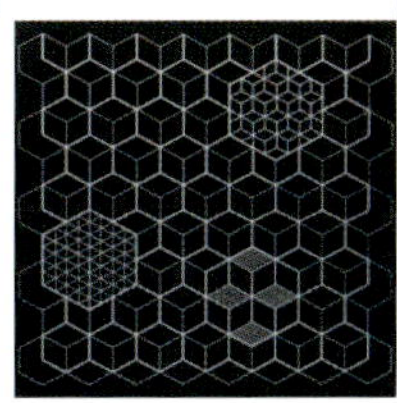
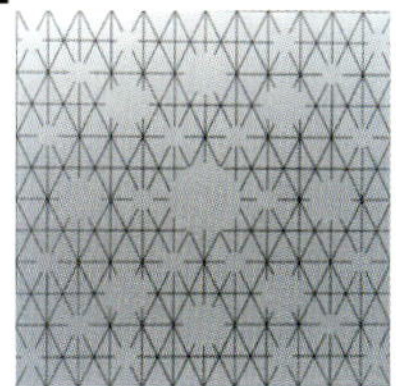
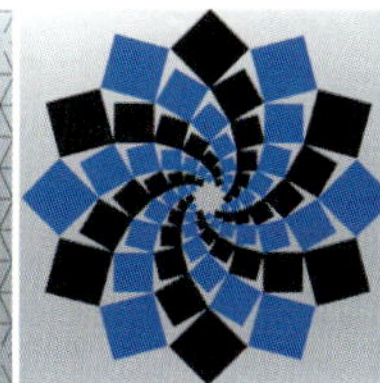
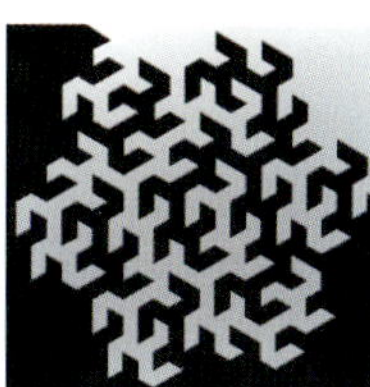

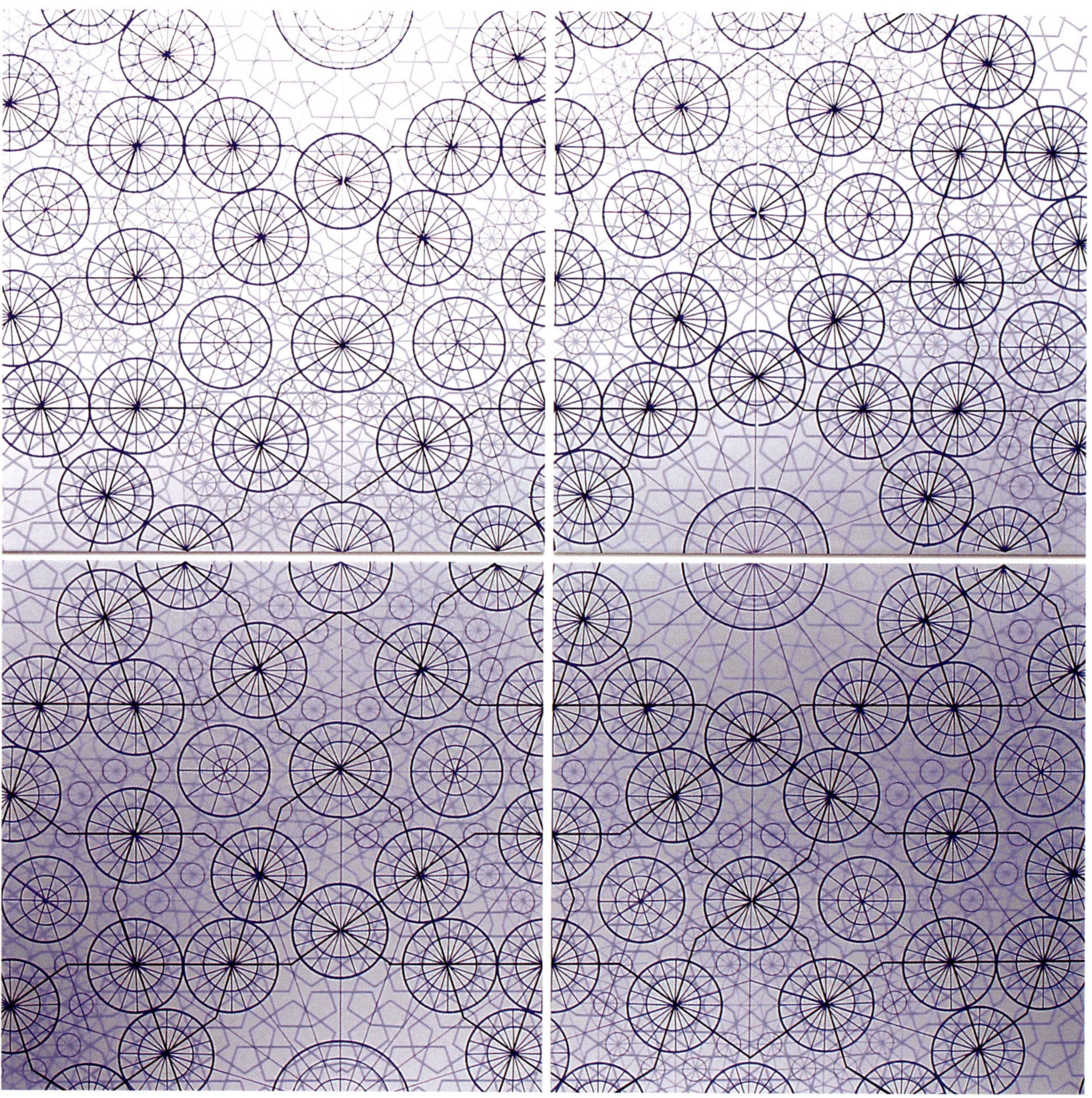

TOPKAPI PALACE STUDY 2002, SCREENED ON ALUMINUM, 36" X 36", FIRST FLOOR CORRIDOR

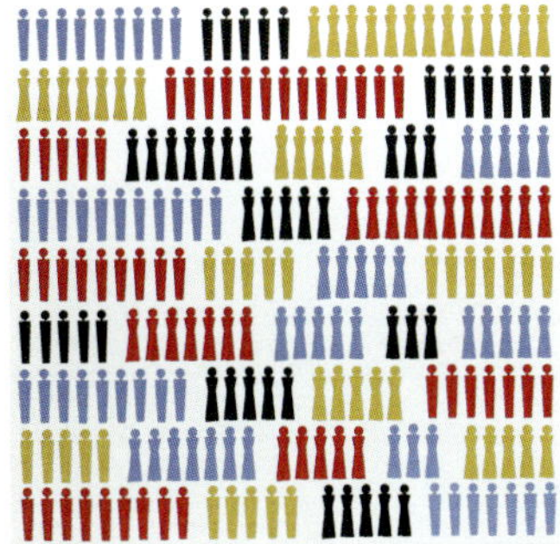

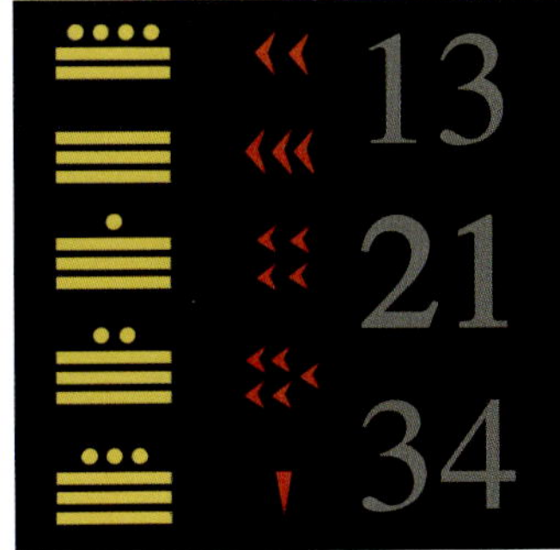

electric waves, moirè patterns, music, Islamic calligraphy, Indian designs, continuing on to outer space.

Bliss looks upon mathematics as basic to how we organize the visual world. Her observations of the structure of natural forms have extended strong influence on both her art and architecture. From her travels and studies she is fascinated by the forms and patterns of many cultures.

So the movement of **Numbers and Measure** begins in the first floor lobby where "Discoverers" meet beneath a fractal tree. She turns to him to offer the red apple of wisdom (pl.1). Beside them the descent of numbers unfolds (pls. 2, 4, 9)

From this stillpoint, clusters carry the viewer to variations of rhythm and measure to the choreography of dance, in the ordered lineup of Census 2000, in the flowing calligraphy of Islam, in the characters of a Chinese triangle, in the words of a sonnet, in the notes from a jazz composition, in the red spiral using Fibonacci proportions to return to the beginning of the dance.

On the second floor **Intersections** is the theme where nature, art and mathematics meet. Here in the lobby one wanders through the geometry of cultivated painted fields, stops beside the rippling surface of a Lake created from randomly generated Truchet tiles and sits on the bank of ferns at the edge of a fractal forest. All is balanced. Japanese screens and Seurat's island of "La Grande Jatte" float through my mind.

For the third floor **Exploring Outer Space**, Anna's statement is the best guide to navigating this realm:

"Mathematics has helped to raise the ceiling of our world from myth and theory to the experience of outer space. Human exploration was made possible by adding precision to computer development and advanced telescopes."

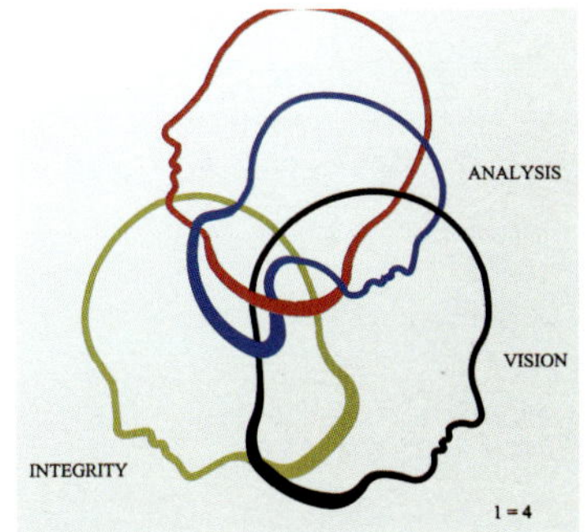

ASPECTS OF MIND

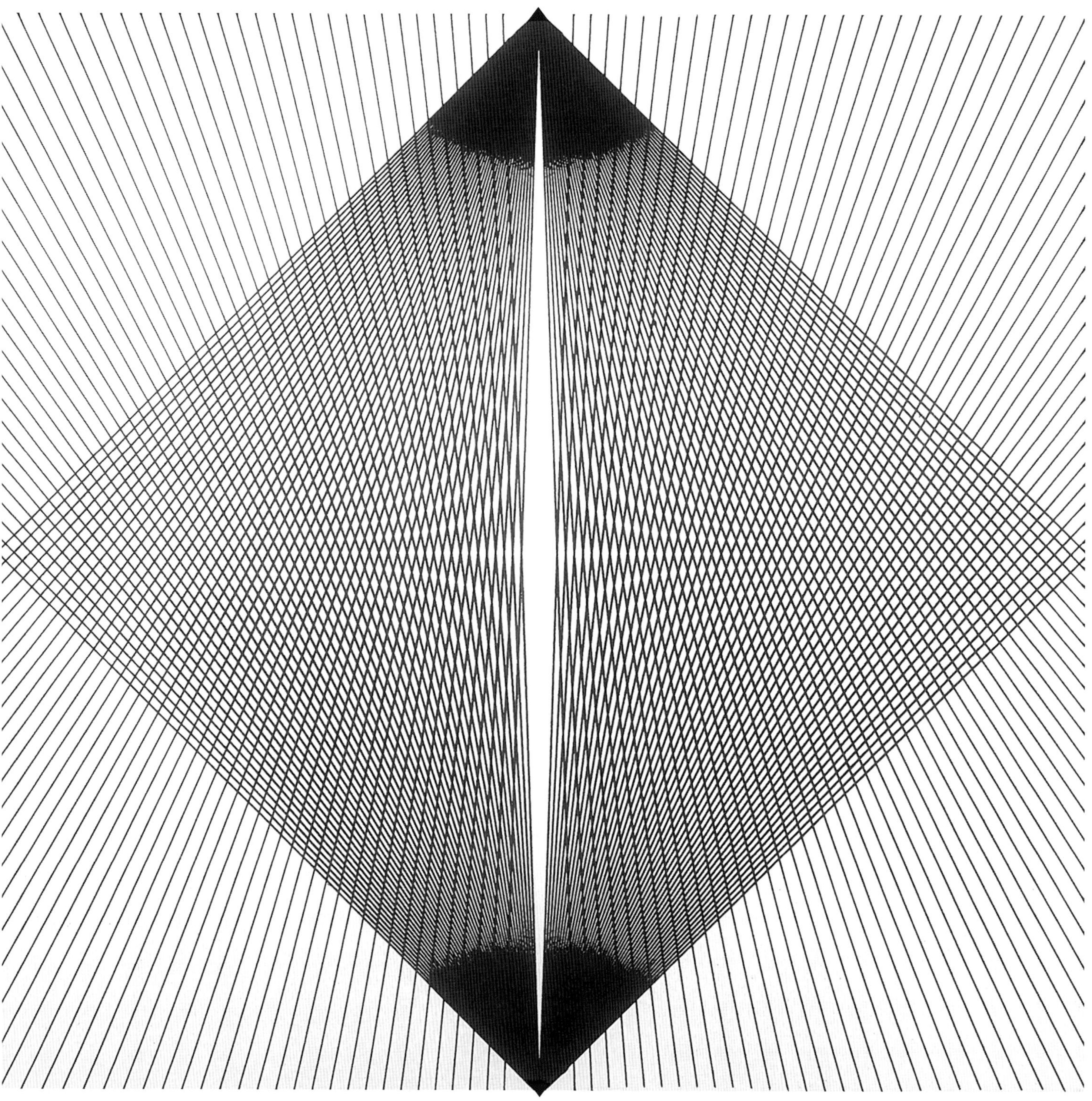

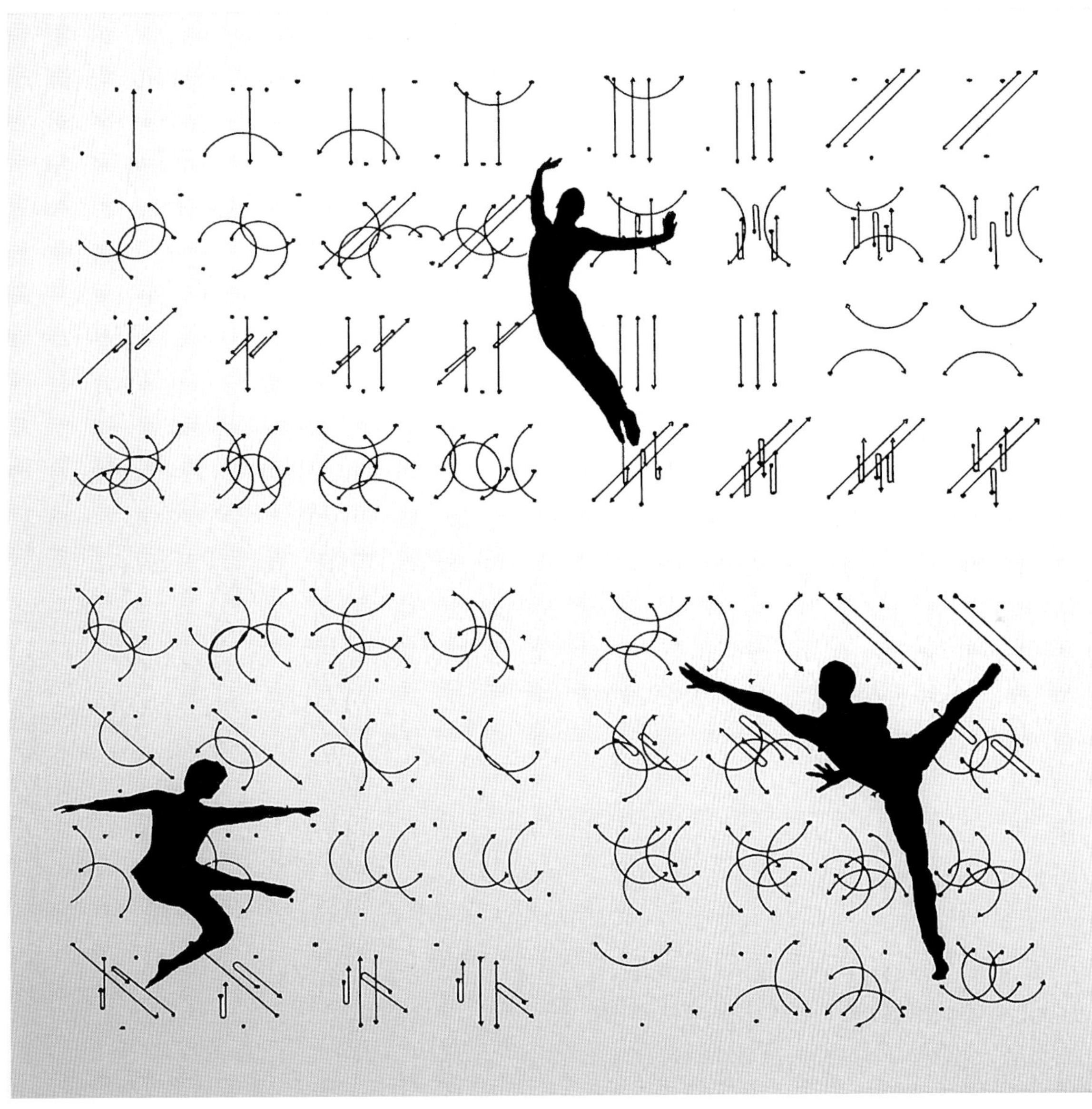

CHOREOGRAPHY BY LUCINDA CHILDS WITH RDT DANCERS & ALVIN AILEY 2001, SCREENED ON ALUMINUM PLATES 18"x 18"

LEFT, **MOIRÈ STUDY** 2001, SCREENED ON ALUMINUM, FIRST FLOOR CORRIDOR

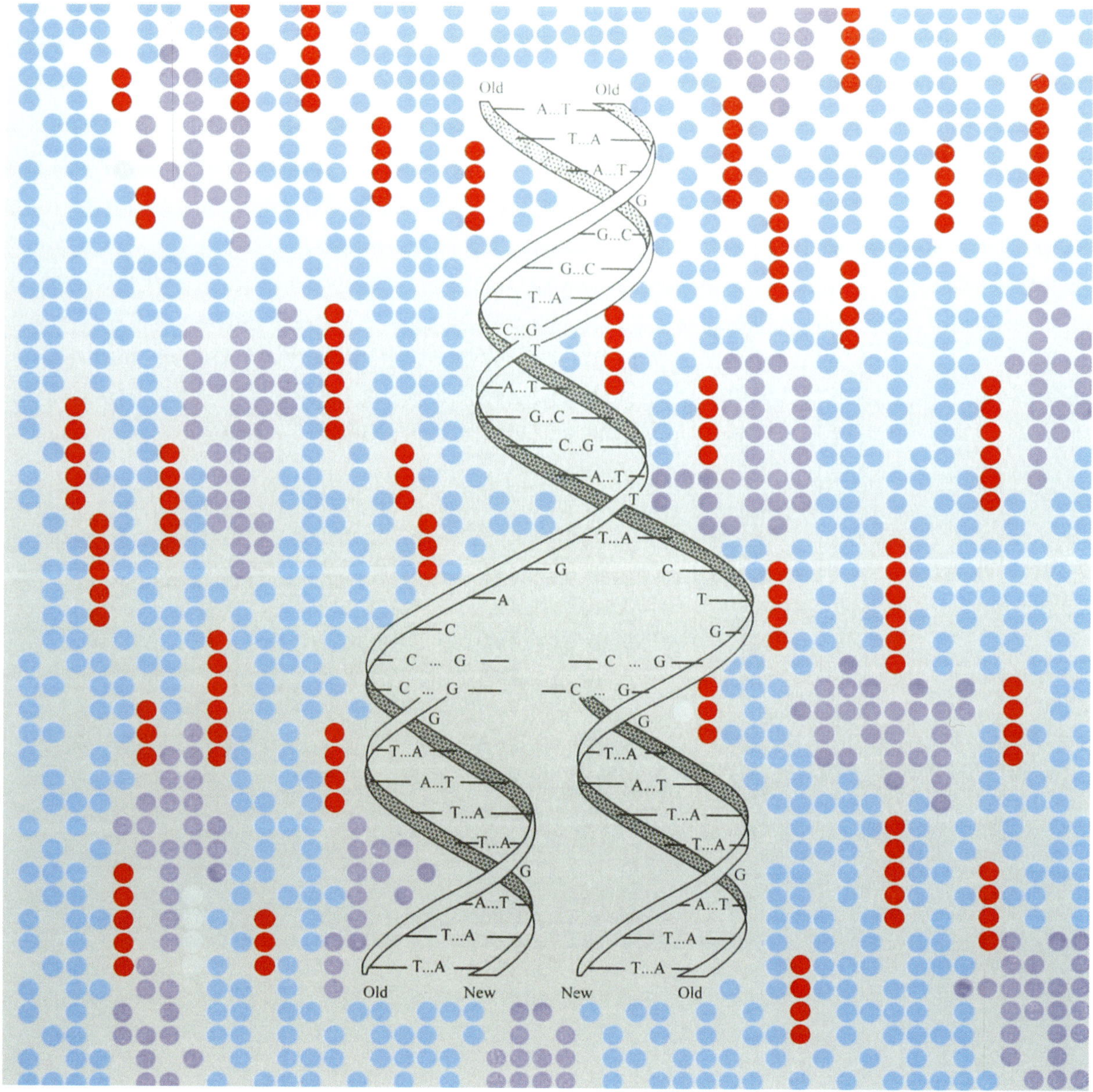

INTERSECTIONS WITH SCIENCE
Screened on aluminum, 18" x 18"
DNA / Microarray 2002

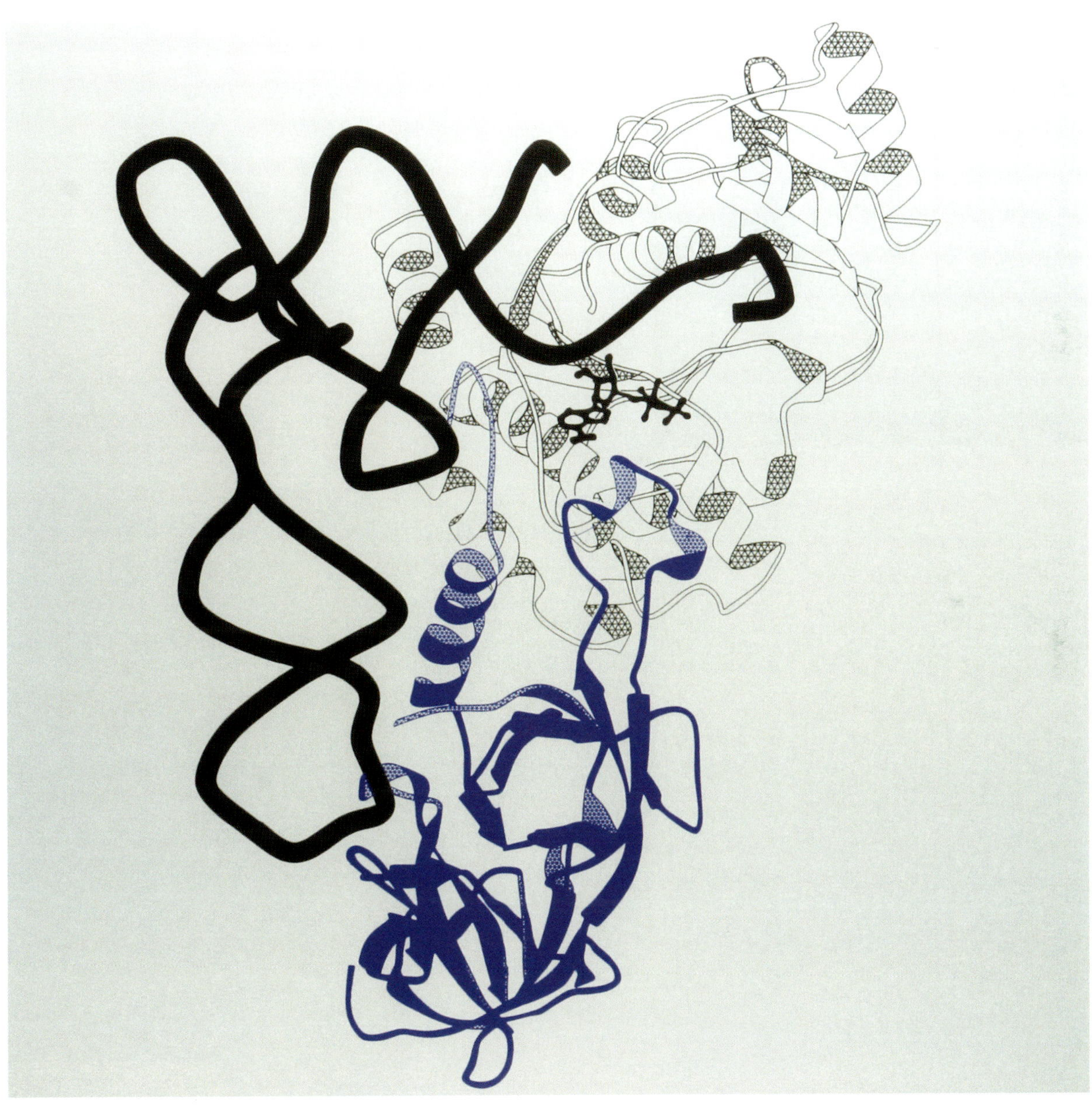

RNA Complex 2002

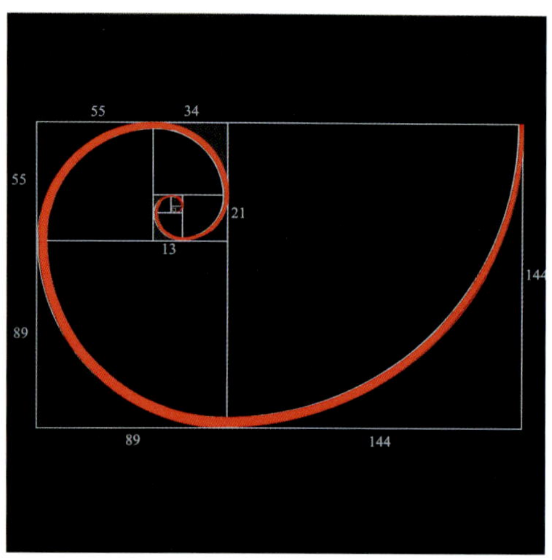

Logarhythmic spiral

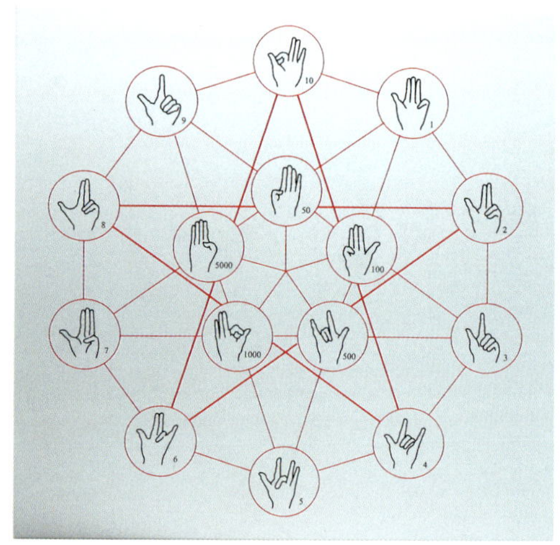

Counting with hands

64

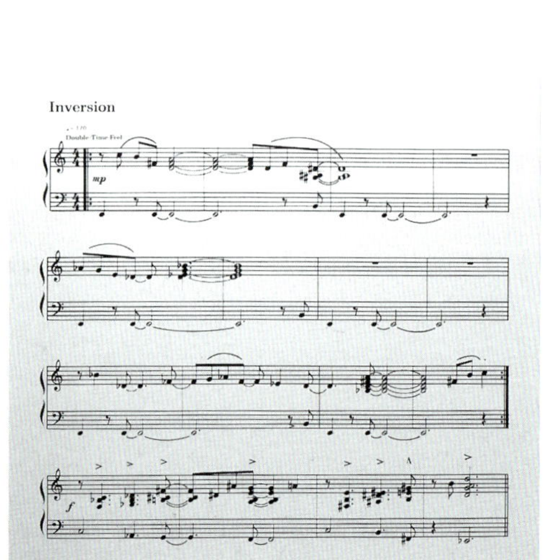

Jazz composition by Dan Waldis, 2001

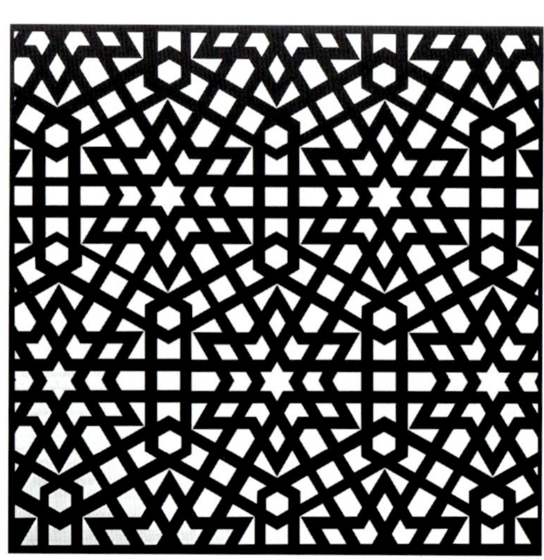

Islamic grill, Central Asia

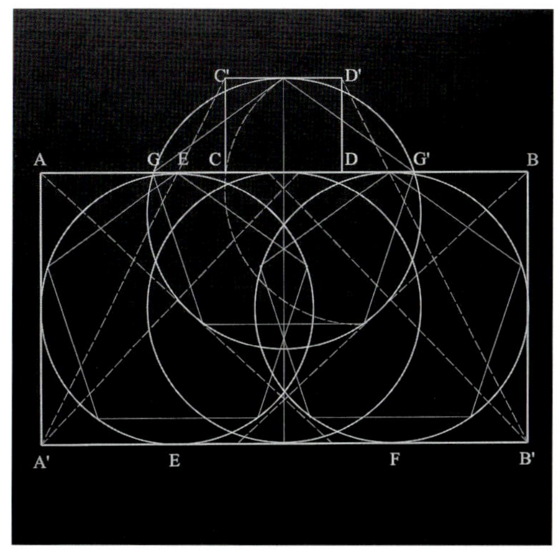

Renaissance concept based
on golden number

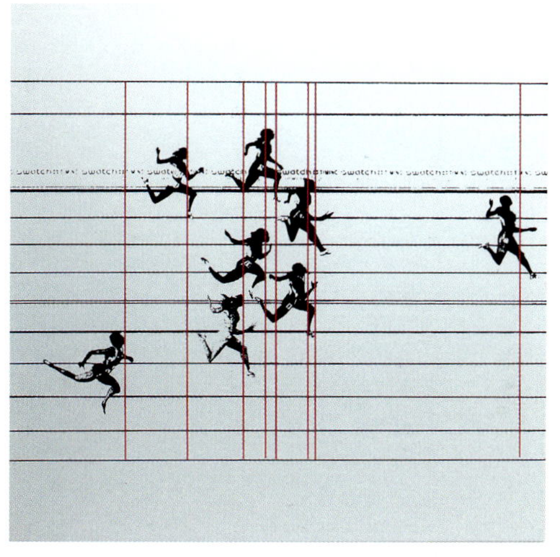

Swatch photo

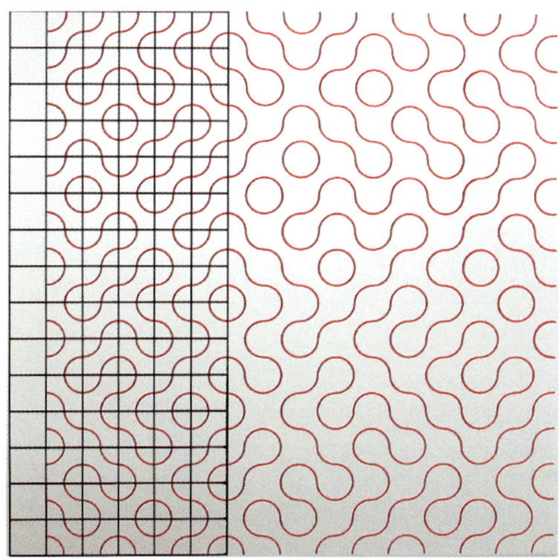

Truchet algorithm tiling

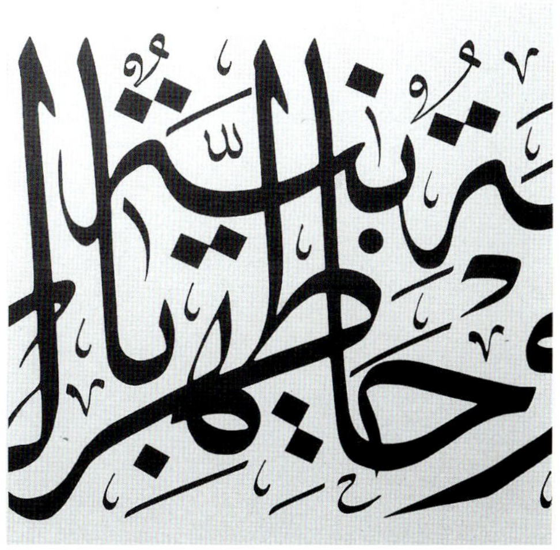

Islamic calligraphy detail

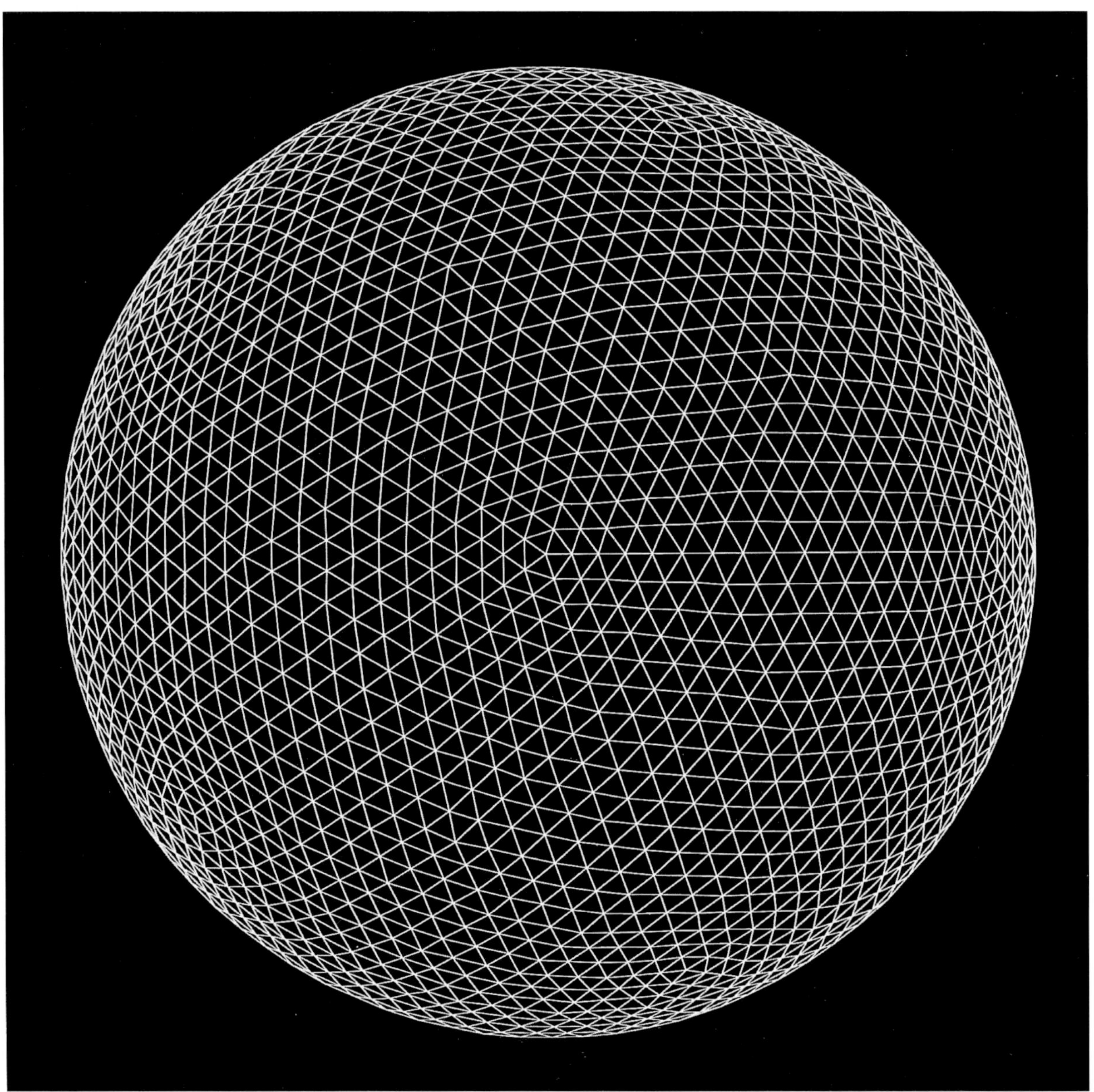

INTERSECTIONS WITH ARCHITECTURE

Montreal Pavilion FOR UNITED STATES, 1967
R. BUCKMINSTER FULLER AND OTHERS

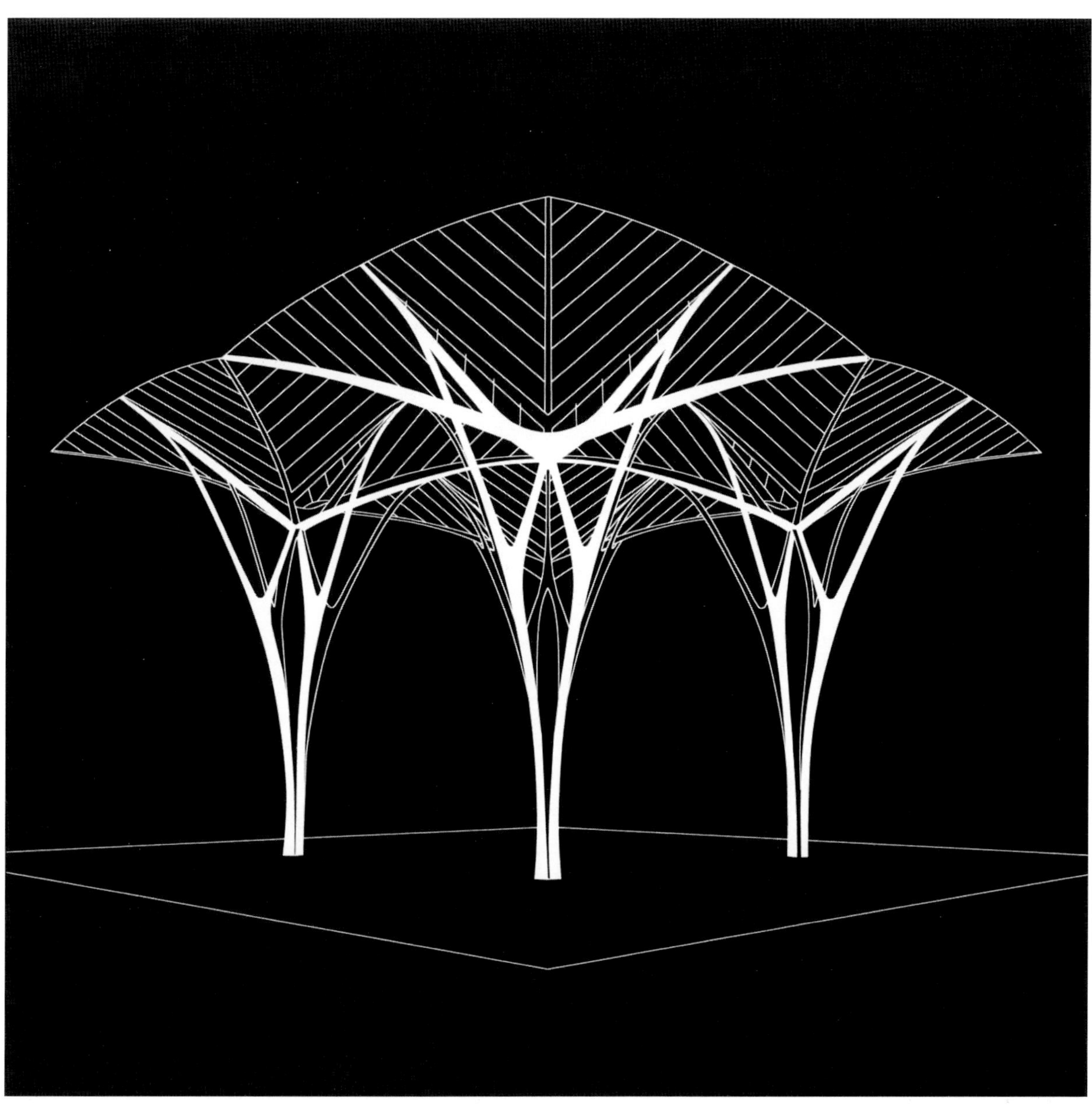

Zurich Pavilion
Santiago Calatrava

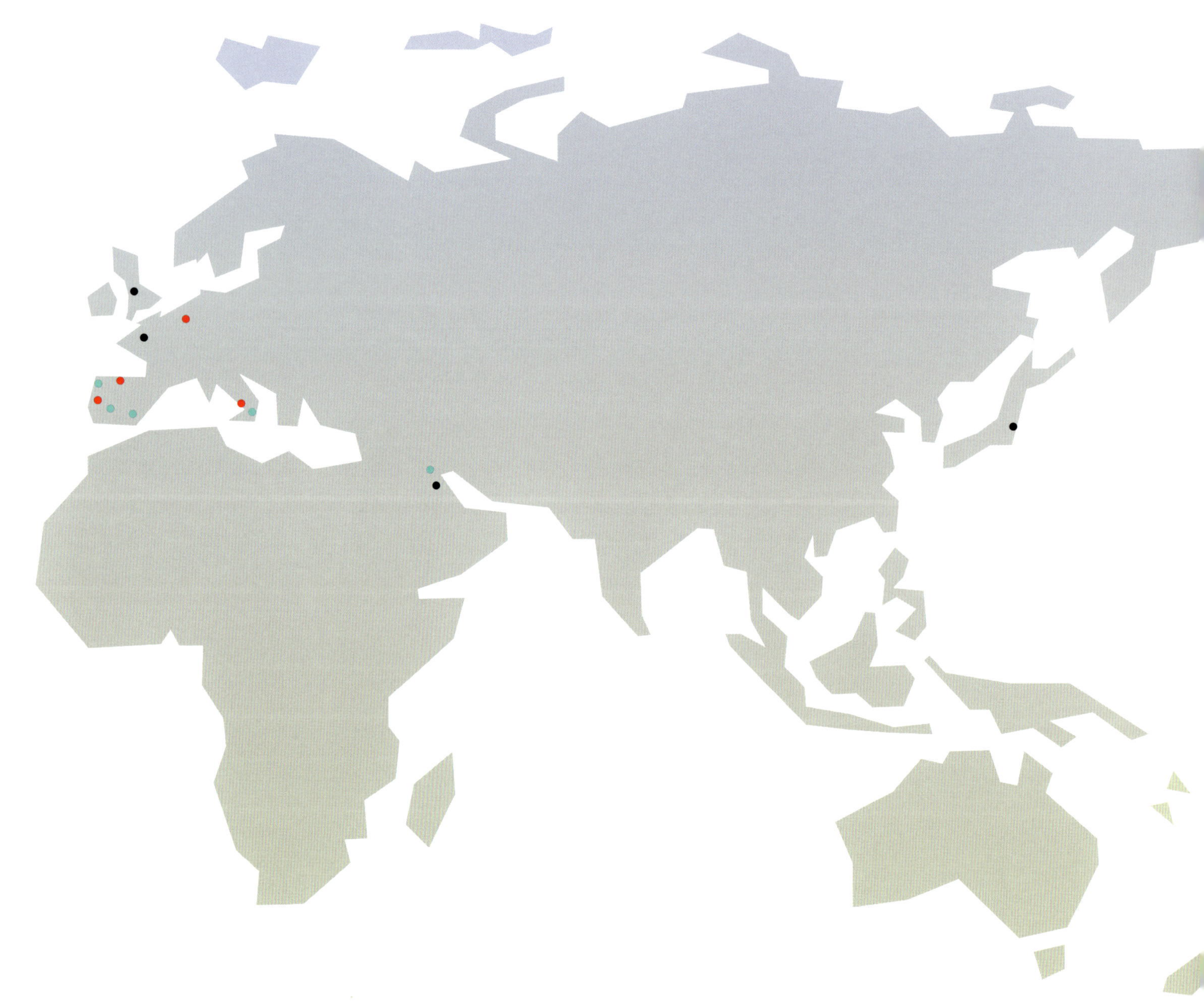

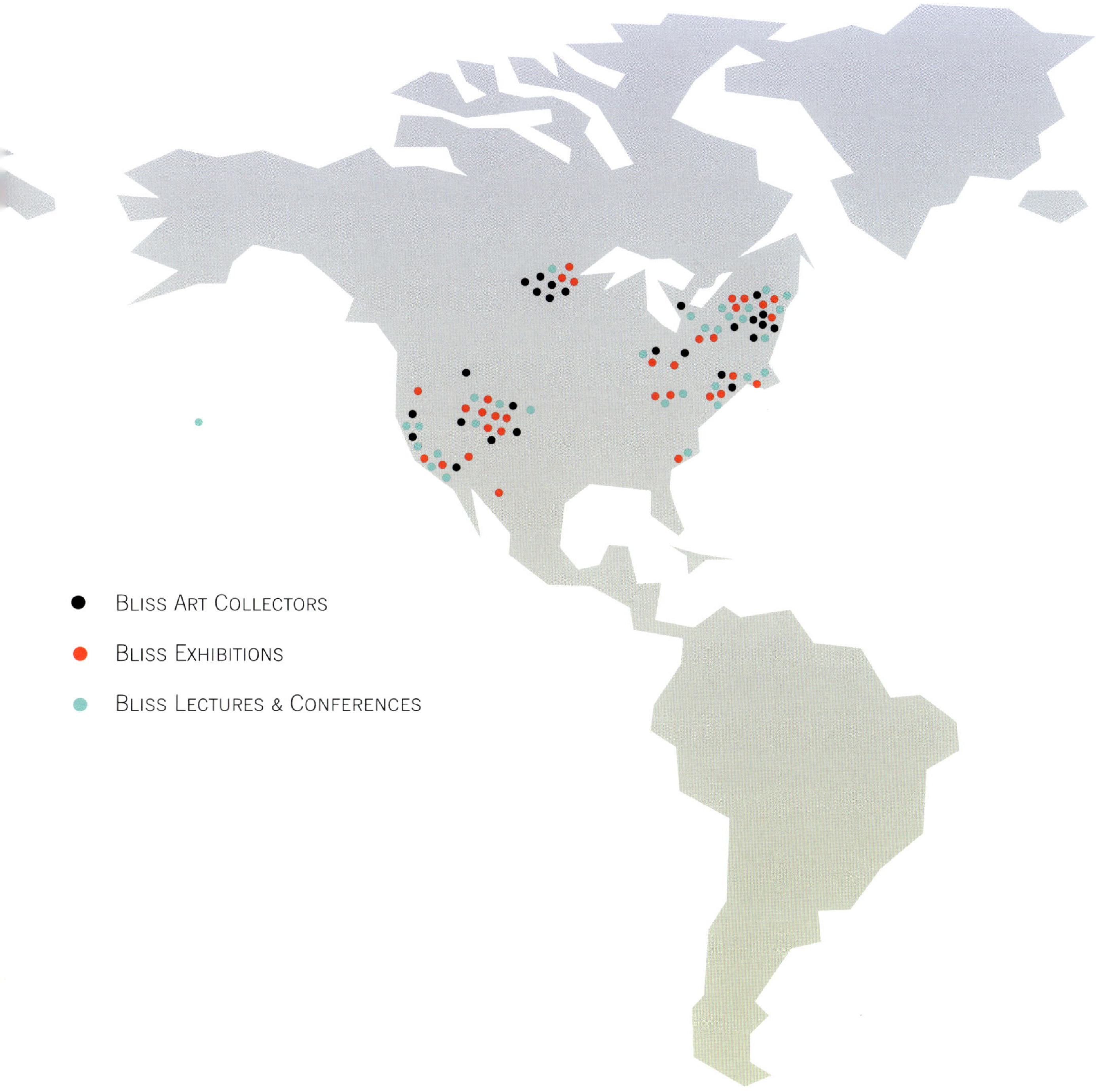

BLISS ART COLLECTORS
BLISS EXHIBITIONS
BLISS LECTURES & CONFERENCES

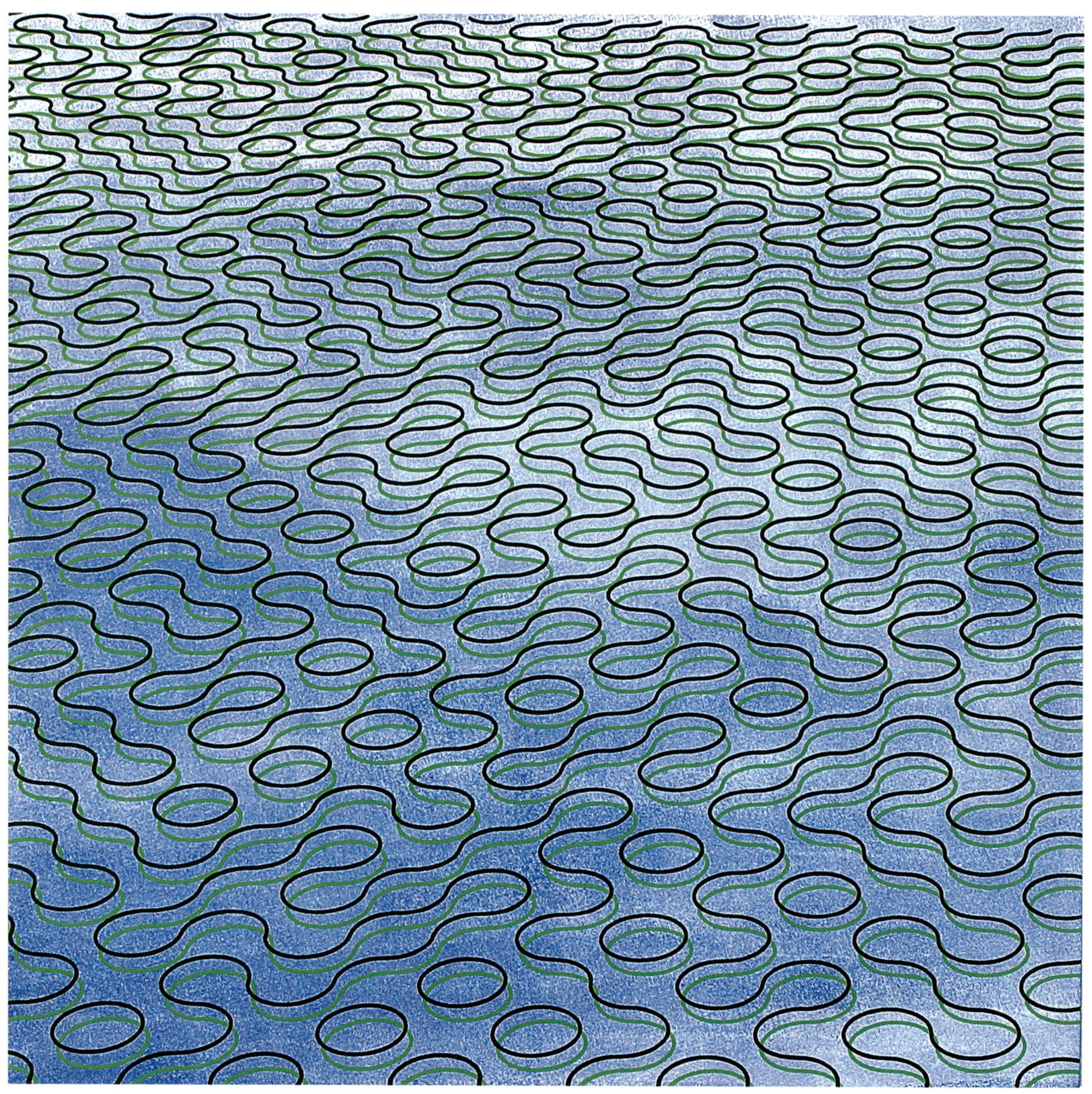

INTERSECTIONS WITH NATURE
Painted & screen printed on aluminum
Computer generated Truchet algorithm

CHRONOLOGY

1925	Born, Morristown, New Jersey
1946	B. Arts Wellesley College
1947-50	Color & Light with Gyorgy Kepes, MIT
1950	B. Arch. Harvard University
1950	Married Robert Lewis Bliss
1950-51	Study Travel in Europe
1954-63	Painting, Drawing & Printing - University of Minnesota & Minneapolis School of Art, 1956 with Josef Albers
1956	Architectural practice, Bliss & Campbell Architects
1963	US Installation at VII Biennial, Sao Paulo. Brazil, Peru and Mexico
1964-74	Computer, Sumie, Languages, Archeology - University of Utah
1975	Moved into remodeled art and color studio
1978	Kuwait, Turkey, Italy

Awards & Honors

1980 Graham Foundation Fellowship for Color Research
1984 Mid-Career Fellowship, American Academy in Rome
Other: Computer World Smithsonian finalist, ASID Presidential Citation,
Nominee for Cooper Hewitt Museum Lifetime Achievement Award,
Painting, Printing & Architectural Awards, Salt Lake City Arts Council &
Utah Arts Council Arts Grants

Civic and Professional

Inter-Society Color Council, Past Board and ASID Chairperson; Founder and Advisor, The Contemporary Arts Group; Former Boards: Repertory Dance Theatre, Salt Lake City Design Board and UAC Committees; Restoration Board for Cathedral of the Madeleine, 1983-90, 1992; Utah Museum of Fine Arts Advisory Board 1972-1998; Member: Ylem, S.F. and Art Science Collaborative, NYC, Utah Women's Forum.

Collections

Metropolitan Museum, New York City	Utah Museum of Fine Arts, Salt Lake City
The Art Institute of Chicago	Salt Lake Art Center, Salt Lake City
Springville Museum of Art	Marriott Library, University of Utah
F.R. Weisman Art Museum, Minneapolis	British Petroleum, (Amoco, Standard Oil) Chicago
Minami Gallery, Tokyo	Utah Innovation Center, Salt Lake City
Salt Lake County	Minnesota Mining & Mfg. Co., St.Paul, MN
Utah State Collection	Detroit Plaza Hotel, Detroit, MI
Novell, Inc. Minneapolis	Prudential Insurance Co., IN
Salt Lake City International Airport	First National Bank, Minneapolis,
Utah State Retirement Board, Salt Lake City	Federal Reserve Bank, Minneapolis
Idaho First National Bank, Boise	Harris Bank, Chicago
Utah Bank & Trust, Salt Lake City	American Tel. & Tel., San Francisco
Cliff Lodge Fine Arts Collection, Alta, UT	Private Collections

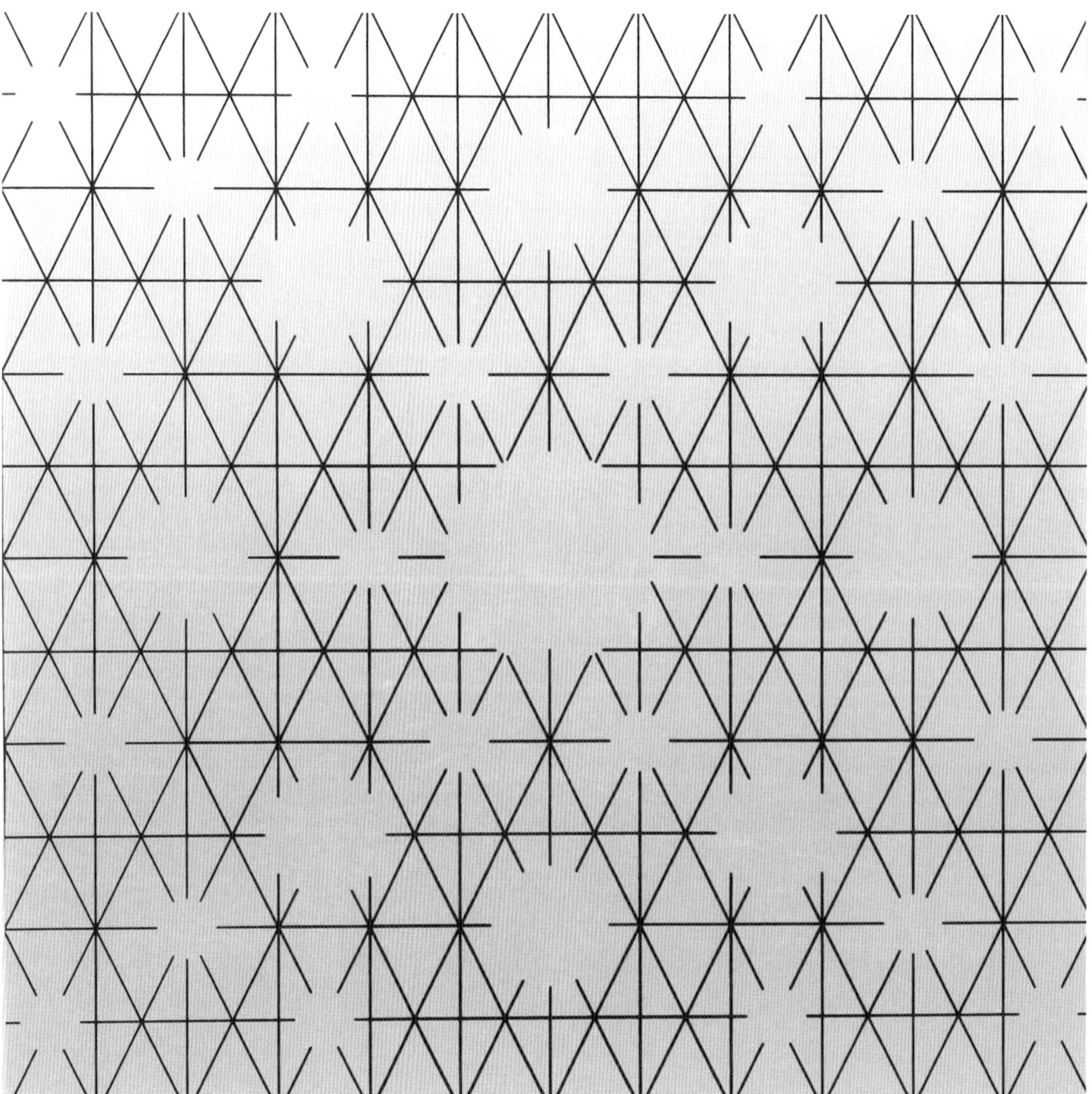

IMPLIED GEOMETRY 2001

A VISUAL ILLUSION, SCREENED ON ALUMINUM

Solo Shows

Utah Museum of Fine Arts, SLC, UT, "INTERSECTIONS: The Art of Anna Campbell Bliss", 2004
Myra Powell Art Gallery, Ogden, UT, "Prints & Projects", 1997
Salt Lake City Arts Council, "Extended Vision", SLC 1993
American Assn. for the Advancement of Science, Washington, DC 1991
Gayle Weyher Gallery, SLC, UT 1987
Salt Lake Art Center, Salt Lake City, UT 1983
Focus Gallery, San Diego Museum of Art, 1981
Atrium Gallery, Salt Lake City Library, 1981
Univ. Gallery of Fine Arts, Ohio State University, 1980
Lee Hall Gallery, Clemson University, Clemson, SC 1979
Dept. of Architecture, Yale University, New Haven, CT 1979
Utah Museum of Fine Arts, Traveling Exhibit, 1979-81
Source Gallery, San Francisco, CA, "Color Into Light", 1977
Lowe Art Gallery, Syracuse, NY, "Color/ Light/Module", 1976
Marriott Library, University of Utah, 1975
Suzanne Kohn Gallery, St. Paul, MN 1974
Salt Lake Art Center, "Modular Serigraphs", 1971

Group Shows
Selected List

"FAAR OUT", American Academy in Rome Fellows Exhibition, ADC Gallery, NewYork, 2003
"ISAMA/BRIDGES Art Exhibition", University of Granada, Spain 2003
"Utah Artists: A 150 Year Survey", Olympics 2002 Cultural Olympiad,Springville Museum
"Juxtapositions: The Artist and the Environment",SLC Arts Council, Finch Lane Gallery, 2002
"Impact of Ylem: 20 Years of Art, Science and Technology", San Francisco, 2001
"Digital '01" ASCI's competition show, New York Hall of Science, Queens, 2001
"Images for a New Millennium", Atrium Gallery, Chapman, SLC Public Library, SLC, UT, 2000
"Westward Bound"- Celebrating Book Arts in the West, Finch Lane Gallery, SLC, UT, 1998
"Folding Screens" - Artists' Invitational, 1999; "Water", Salt Lake Art Center, SLC, UT 1998
"The Reality of Abstraction", 1946 - 96, Nora Eccles Harrison Museum of Art, Logan, UT
        Salt Lake Art Center; Museum of Art, Brigham Young University, 1996-97
 Utah Arts Festival,2001, 1996, 90, 89; "20-20 Vision", 1986, Utah Arts Council 1998
"Utah Women Artists", AAUW Exhibition 2000, Springville Museum, Salt Lake Arts Center, 1997
"The Scientific Tradition in Art", Amer. Assn. for the Advancement of Science, Washington, DC 1996
 Retrospective Exhibition, Jewett Art Center, Wellesley MA, 1996, Alumnae Art Exhibition, 1991
 Art, Xerox PARC Research Lab, Palo Alto, CA 1994
 7th National Computer Art Invitational and Traveling Show, East. Washington Univ. Gallery, 1993 -96
 Utah Painting/Prints, Utah Museum of Fine Arts, SLC 1998,94, 88, 86, 80, 78, 64, Purchase Awards
 Utah Arts Council Visual Artist Exhibition, "A View of Nine", Salt Lake Art Center, 1992
 Computer World Smithsonian Awards Exhibitions, National Building Museum, Washington, DC 1991
"Contemporary Utah Artist", SLC Arts Council, 1990, 86, 80
 Springville Museum of Art, 1993, 91, 83, 81, 80
"An Elegant Merging" (Computer & Art), Salt Lake Art Center 1989, Kansas City Art Institute, 1988
"Western State Print Competiton", Eccles Com. Art Center, 1988
"Contemporary Women Artists", UMFA, 1987, 85, 83, 76 Awards
 International Color/Design Competition, Stuttgart Design Center, Germany, 1986-87, 1980-81
"Computer as Art", The Science Place, Dallas, TX 1986
"Critics Choice", Utah Museum of Fine Arts, SLC, 1985
"Art for Worship", Salt Lake Arts Council, SLC, 1985
 American Academy in Rome Annual Exhibition, 1984
"Printmakers West", N.E. Harrison Museum, Logan , UT 1983
"Color/Light/Architecture", Graham Foundation, Chicago, 1983
"Eleven Utah Printmakers", Arizona 1982
 Museum of Fine Arts, Richmond, VA 1980
"Utah Originals", Utah Arts Council Exhibit, 1979
"Four Women Artists", Kimball Art Center, Park City, UT 1978

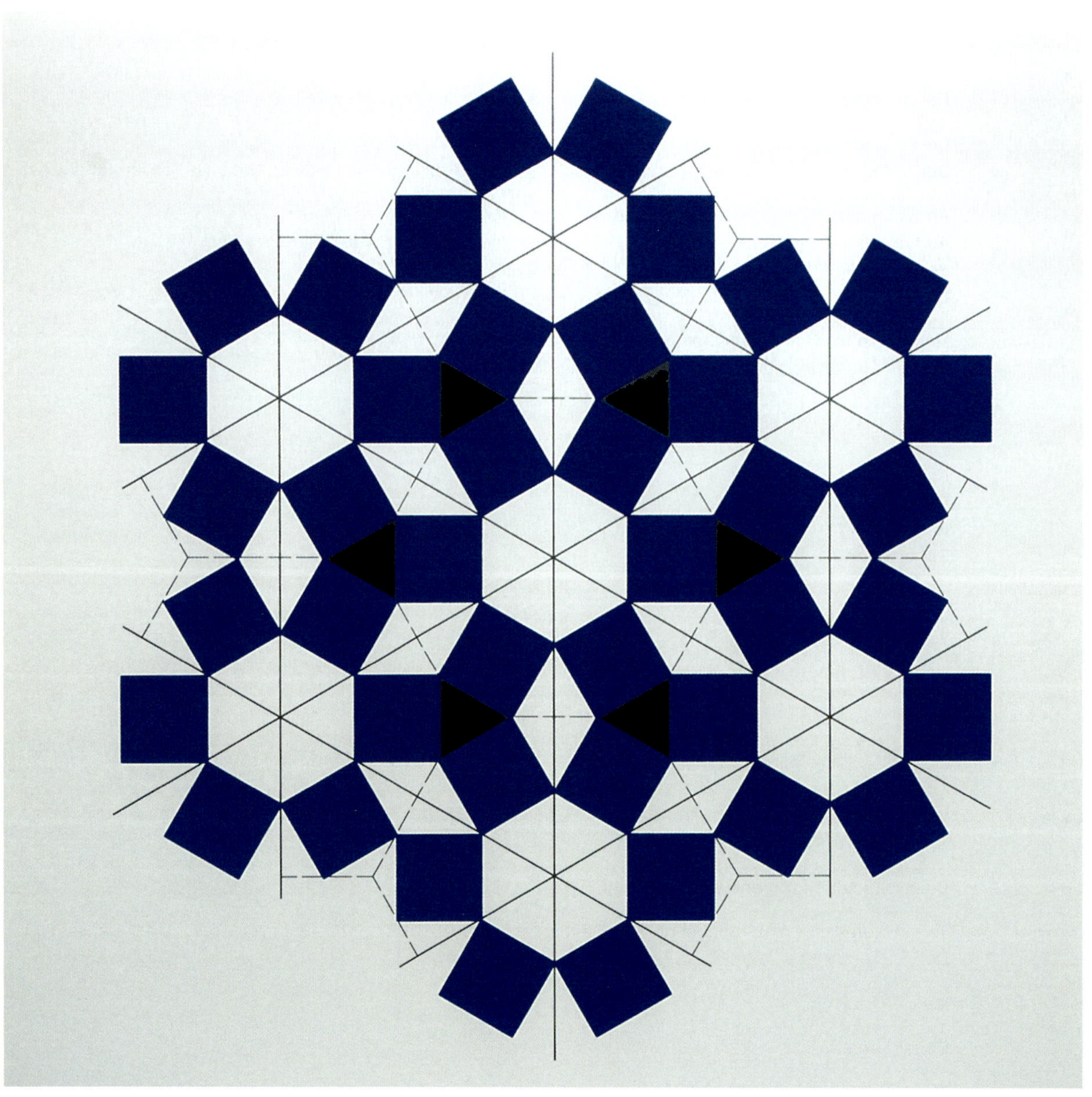

ISLAMIC PATTERN
Screened on aluminum, 2001

**LECTURES &
SEMINARS**

**Inter-Society Color
Council:**

2nd Panchromatic Conference, 2000 "Moving Color into Three Dimensional Space";
1998 Williamsburg Conference: "Color & Design", Co-Chair, Poster Exhibit; "WINDOWS', Poster & Video, Cleveland 1990,
Poster, Detroit 1994; Color in Architecture Symposium", Co - Chair., Chicago 1989
Research Reports 1994, 1990, 1989, 1986, 1977; "Off Color Learning", 50th Annual Meeting, NYC 1977

**American Institute of
Architects Seminars:**

"Color in Architecture", AIA Western International SUMMIT 2000,
Sun Valley Idaho; AIA state meeting 1997, Sun Valley; "Color in Architecture", National Convention, Hawaii 1982

Mathematics & Art:

"An Extended Mural for a House of Mathematics" ISAMA/BRIDGES 2003, Univ. of Granada,
"Architectural Extents", ASAMA 1999, University of the Basque Country, San Sebastian, Spain.
Seminar, Westminster College, SLC 1994; "Between Art and Architecture", AM 93, SUNY, Albany
"Extended Vision", SLC Arts Council gallery talks, 1993; "INTERSECTIONS", Utah Math/Science Network, SLC 1993

**National Computer
Graphics Association:**

"Revolution in the Arts", Anaheim, CA 1990

**MIT Center for
Advanced Visual Studies:**

"Art Transition '90", Artist's Presentation & Session Chair.

**Women's Caucus
for Art:**

"On Being An Artist Today", San Francisco, 1989

University of Utah:

"Roots of Modern Architecture: The Climate for Innovation", Grad. School of Arch. 1995
Visiting Lecturer on Color, GSA '88, '69, '67; "Color & Ornament," Graduate . School of Architecture 1985
"Is There an Antidote to Academic Color?" Arts 1983; "Color and Illusion", University Lecture 1973
"Spanish Influence on Contemporary American Architecture" KUED 1969

**American Academy
in Rome:**

"Color in Art & Architecture," 1987; "Concepts & Contexts", 1984

Graham Foundation

"Color/Light/Architecture", Chicago, 1983

**American Society of
Interior Designers:**

CEU Seminars, 1988, 1987 1986; Tenn. State Mtg., Memphis 1991
"Color Resources for Interior Designers", National Conf., Boston 1983; "New Perspectives in Color," Regional Conference, SLC 1981
"Visual and Psychological Aspects of Color Important for the Designer," National Conference, Los Angeles 1976
"Walls," National Conference, Denver 1974

Color Marketing Group:

"What Color is a City?", National Conference, Washington, DC 1983
"The Design of Fantasy", Regional Conf., Las Vegas 1981;"Bedside Color" (Hospitals), National Conf. Nashville 1980

Wool Bureau Conf.

"Color, the Added Dimension," Atlanta 1983

Salt Lake Art Center

"Ten Lectures on Color," School of Art , Gallery talks: 1998,1996, 1992 About My Work, 1991 Issues in Art
Docent Lectures: 1981 - 83 ,1978 ,1974, 1972, 1971 ,1965 - 67

Utah Arts Council:

Visual Arts Panel, 1990, "The Artist's Image," 1982

Wellesley College:

"Contemporary Architecture" Careers Conference 1972

Walker Art Center

"The Art of Interiors" 1963, Minneapolis

**Visiting Lecturer
on Color:**

Univ. of Virginia 1982, Clemson Univ. 1979, Syracuse Univ. 1976
Ohio State Univ. 1980, U. of Calif., LA 1977, Univ. of Maryland 1976
Yale Univ. 1979, Kuwait Univ. 1978, Utah State University 1975
Cal Poly State University, San Luis Obispo 1975, 1973, 1970

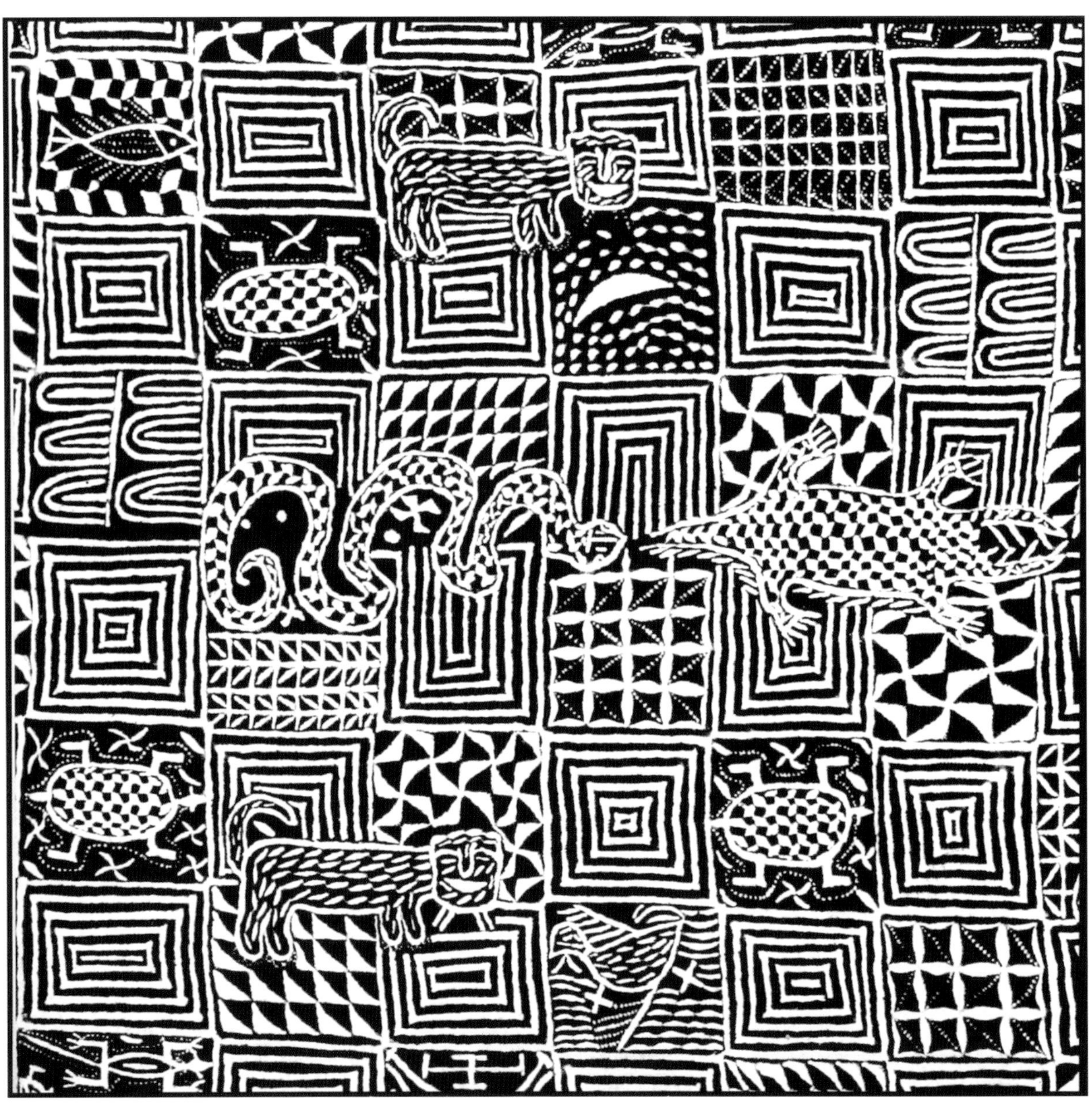

NIGERIAN TEXTILE DESIGN, resist dyed
LASER ETCHED ON BLACK ANODIZED ALUMINUM

BIBLIOGRAPHY

**Publications by
Anna Campbell Bliss**

"An Extended Mural for a House of Mathematics",Conference paper, "Meeting Alhambra", ISAMA/BRIDGES
    proceedings, edited by Barrallo, Friedman, Sarhangi, Sequin, Univ.of Granada, Spain, July 2003, pp 273-282
"Architectural Extents", Conference paper, ASAMI 99 proceedings ,edited by N. Friedman & J. Barrallo.
    San Sebastian, Spain, June 1999, pp. 61-65.
"Labyrinths of the Mind", artist's book, Salt Lake City 1998, revised edition 2000..
"Explorers", The Leonardo Gallery, LEONARDO, Vol. 28, No. 4, 1994, pp. 239-242.
"Mathematics for the Garden of the Mind", LEONARDO, Vol. 26, No. 1, 1993, pp. 19-22.
Review of the Second Computer Revolution: Visualization by Richard Mark Friedhoff, COLOR RESEARCH
    AND APPLICATION, Vol. 18, No. 2, April 1993.
Review of Primary Sources, Selected Writings on Color from Aristotle to Albers by Patricia Sloane,
    COLOR RESEARCH AND APPLICATION, 1992.
"Art Criticism?", SALT LAKE ART CENTER, Spring 1991, pp. 1-2.
"Mixed Media Studies in Tactility: An Alternate to Computer Art", LEONARDO, Supplement, 1988, p 117.
"Talking About Color: Art and Visual Measure", COLOR RESEARCH AND APPLICATION, Vol. 13, No. 6,
    Dec. 1988, p 345.
"New Technologies of Art-Where Art and Science Meet", LEONARDO, Vol. 19, No. 4, 1986, pp. 311-316.
"New Technologies of Art", UTAH ARCHITECT, Fall 1985.
"Art, Color, Architecture", AIA JOURNAL, Vol. 71, No. 2, Feb. 1982, pp. 48-55.
"Color for the Hospital Environment," THE DESIGNER, Nov. 1982, pp. 26, 55.
Review of Color and Human Response By Faber Birren, COLOR RESEARCH AND APPLICATION, Vol. 5,
    No. 3, Fall 1980, pp. 183-84.
"Color Selection as a Design Decision", AIA JOURNAL, Vol. 67, No. 12, Oct. 1978, pp. 60-65.
"Color/Light/Module", Screen Prints, Salt Lake City, Aug. 1973.
"Ambivalent Color", UTAH ARCHITECT, Summer 1967.
"Children's Furniture", Guest Editor. DESIGN QUARTERLY 57. 1963.

**Articles and
Reviews on Art**

Roper, Roger, "Artists in Residence (historic buildings): Anna Campbell Bliss, UTAH PRESERVATION, vol. 6, 2002
Gagon, Dave, "Art That Adds Up," DESERET NEWS, May 5, 2002, pp E1-2
Gagon, Dave, "The Art of BLISS," DESERET NEWS, Feb. 28, 1999, pp E1-2
McEntire, Frank, "Pixel Perfect": Kenvin Lyman and Anna Campbell Bliss, SALT LAKE, March/April 1999, pp 35-38
"Anna Campbell Bliss", YLEM, February 1994.
Brown-Wagner, Alice, "Review of Extended Vision" SALT LAKE TRIBUNE, Oct. 31, 1993.
Johnson, Jennifer, "Utah's Emerging Women," SALT LAKE CITY, Jan/Feb 1993, pp56-66, 105
Donohoe, Anne Palmer, "Painted Spaces", SALT LAKE CITY, July/Aug. 1992, pp. 88-89.
Poore, Ann, "Utahn's Computer-based Mural Makes Impression on High-tech World",
    SALT LAKE TRIBUNE, May 5, 1991.
Dertouzos, Michael L., "Communications, Computers and Networks," WINDOWS: a Computer Based Mural
    illustrated, SCIENTIFIC AMERICAN, September 1991, pp 32-33
"Detail of WINDOWS", illustrated, SCIENCE, Vol. 251, 22, Feb. 1991, p 953.
Dibble, George, "Anna Campbell Bliss Explores Uncharted Territory in Mural",
    SALT LAKE TRIBUNE, March 25, 1990.
Christenson, Richard, "Review", DESERET NEWS, Sept. 27, 1987.
Christenson, Richard, "Review", DESERET NEWS, Jan. 30, 1983.
"Review" & "News", SALT LAKE TRIBUNE, Jan. 2, 1983, & Jan 16, 1983.

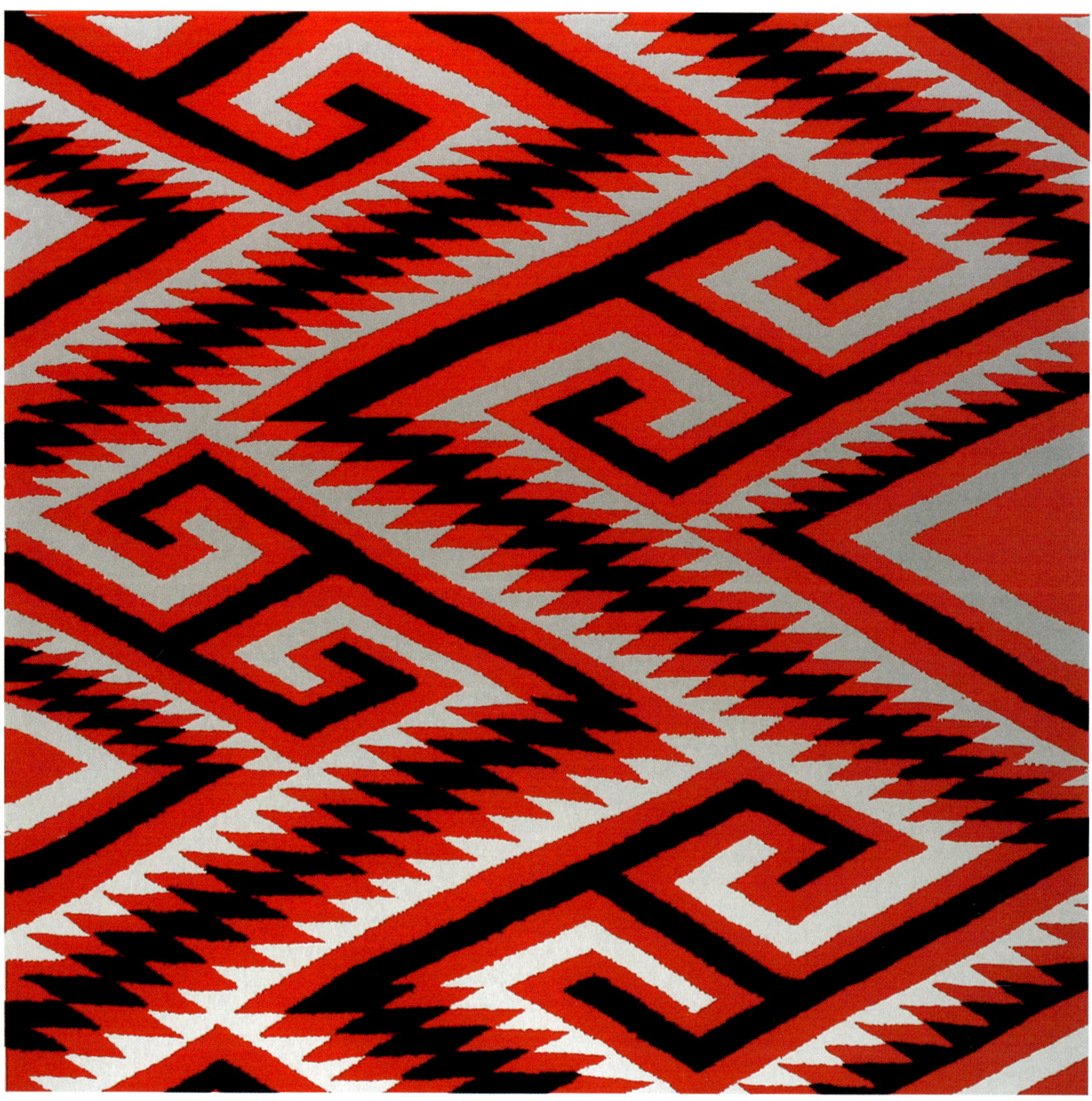

INTERSECTIONS WITH THE ARTS
NAVAJO WEAVING, SCREEN PRINTED ON ALUMINUM

Dibble, George, "Exploring the Fascination of Color", SALT LAKE TRIBUNE, Feb. 15, 1981.
Nelson, Katherine, "Art/Dimensions of Design", UTAH HOLIDAY, April 1983, p 18.
Groseclose, Barbara, "Color in the Art of Anna Campbell Bliss", DIALOGUE, March/April 1980.
Ferrara, Dan, "Art Exhibit Focuses on Color"' OHIO STATE LANTERN, April 23, 1980.
Lanyi, Eva, "Color/Light Exhibition Lets Viewer Perceive Meanings, Pleasures of Color",
    SYRACUSE UNIVERSITY DAILY ORANGE, Sept. 24, 1976.
Olpin, Robert S., "Contemporary Utah Artists Exhibition", UTAH HOLIDAY, Vol. III, No. 16, 1974, pp. 15-22.
.Morrison, Don, "Review," MINNEAPOLIS STAR, Oct. 24, 1974
Flanagan, Barbara, "News & Comments", MINNEAPOLIS STAR, Oct. 10, 1974.
Dibble, George, "Review", SALT LAKE TRIBUNE, Aug. 29, 1971.
Hartranft, Ann, "Review", SYRACUSE HERALD AMERICAN, Oct. 17, 1976.
Appelhof, Ruth Ann, "Review", SYRACUSE NEW TIMES, Nov. 6, 1976.
"M.I.T. Explores Sight and Structure" in "Visual Education of Architects", ARCHITECTURAL FORUM,
    Vol. 92, March 1950, p. 152.

Books and Major Catalogs

"Utah Artists: A 150 Year Survey", Olympics 2002 "Cultural Olympiad", Springville Museum
"Westward Bound" catalog of traveling exhibition celebrating Book Arts in the West, Special Collections,
    J. Williard Marriott Library, University of Utah.
"The Reality of Abstraction: Painting in Utah 1946-1996," Ann Poore, cur., Nora Eccles Harrison Museum of Art.
Swanson, V.G.; Olpin, R.S.; Seifrit, W.C., UTAH ART, Layton, UT, Peregrine Smith Books, 1991, rev. 1997.
"A View of Nine", Utah Arts Council Exhibition, Salt Lake Arts Center, 1992.
"A Search for Heroes", Computerworld Smithsonian Awards, 1991.
"Making and Breaking Tradition - 60 Years", Salt Lake Art Center, 1991.
"First International Colour Design Prize 1980/81", Stuttgart Design Center, Germany.
"Third International Colour Design Prize 1986/87", Stuttgart Design Center, Germany.
"Salt Lake County Fine Arts Collection", Salt Lake City, UT, 1987.
"Annual Exhibition American Academy in Rome", June 4-5, 1984.
"66th Utah Spring Salon", Springville Museum of Art, March 31-May 20, 1990.
"20-20 Vision", Katherine Nelson, ed. Utah Arts Festival, 1986.
"Women Artists of Utah", Springville Museum of Art, Nov. 3-Dec. 17, 1984.
"Utah Originals", Utah Arts Council, 1979.

World Wide Web Presentations:

Digital Sites: NYC/LA August 6 - Sept 1995, BI-COASTAL ART OPENING art/computer interface
    Mark Frisk & Cynthia Pannuci / ASCI hosts

Architecture and Interiors

Freed, Stacey, "When Design Compels Movement," GARDEN DESIGN, Vol. 5, No. 3, 1986 pp 64-69
"A Beautifully Planned Kitchen", BETTER HOMES AND GARDENS Improvement Ideas, Fall/Winter, 1974-75, pp. 94-95.
"Designing a Kitchen", BETTER HOMES AND GARDENS Kitchen and Family Room Ideas, 1975, pp. 46-47.
"25 Prize Buildings for the Community", ARCHITECTURAL FORUM, Vol. 104, No. 3, March 1956, p. 139.
"Record Houses of 1963", ARCHITECTURAL RECORD, Vol. 133, No. 6, Mid May, 1965, pp. 84-87.
"New Talent Annual", ART IN AMERICA, Vol. 47, No. 1 Spring 1959.
"Celanese Interior Design Competition for Young Professionals", INTERIORS, Vol. 121, No. 4, Nov. 1961, p. 36.
"Bliss and Campbell Architects", NORTHWEST ARCHITECT, Vol. 27, No. 2, 1963, pp. 19-25.

Lenders to the exhibition:

Louis and Magda Jakovcev Ulrich

Dave Reybould

William and Betsy Campbell

Anna Campbell Bliss

Michael Adamson

British Petroleum Corp.

Utah Arts Council

Utah Museum of Fine Arts

Debra Anderson

Mathew Bousquette

Peter and Hon. Judy Atherton

Roger and Jeen Brown

John R. and Sue Smith

Noel and Klancy DeNevers

Edgar John Davies

Janet Minden

Rita Fordham

Ted and Yeiko Nagata

Photographic credits:

Skylar Nielsen

Jim Frankowski

Pat King

Scott Zimmerman

John Telford

Anna Campbell Bliss

Jan Stevenson

Busath

Bri Alva Knorr